INTRODUCTION TO GEOMETRY

Modules in Technical Mathematics

INTRODUCTION TO GEOMETRY

by *Allyn J. Washington*

Dutchess Community College
Poughkeepsie, New York

Copyright © 1976 by Cummings Publishing Company, Inc.
Philippines Copyright 1976

All rights reserved. No part of this publication may be reproduced, stored in a retrieval system, or transmitted, in any form or by any means, electronic, mechanical, photocopying, recording, or otherwise, without the prior written permission of the publisher.
Printed in the United States of America.
Published simultaneously in Canada.
Library of Congress Catalog Card Number 75-27737

ISBN: 0-8465-8616-9

CDEFGHIJK—AL—7987

Cummings Publishing Company, Inc.
2727 Sand Hill Road
Menlo Park, California 94025

PREFACE

Introduction to Geometry is intended for those students who require an introduction to the basic figures and concepts of geometry. It includes an introduction to the triangle, quadrilateral, and circle, along with the measures of perimeter and area. Also included are the Pythagorean theorem, similar triangles, and an introduction to basic solid geometric figures. The material is self-contained and may be used alone or in conjunction with the other modules in this series of Modules in Technical Mathematics.

This geometry module may be used as a supplement or as an integral part of a basic mathematics course. It is intended primarily for students in technical institutes, vocational schools, and regular or developmental programs in two- and four-year colleges. Although the examples and exercises stress technical applications, the material is appropriate for any student who may require an introduction to the subject of geometry.

It is assumed that the reader has a reasonable understanding of the basic concepts and operations of arithmetic, and of algebraic notation along with certain basic algebraic operations. If the reader does not have this background, it is recommended that a review of the necessary topics be made before a study of this module is started. This material may be found in the modules, *Topics from Arithmetic* and *Introduction to Algebra.*

Each basic concept and operation is outlined and set off. It is then carefully exampled. Along with each example there is a brief set of exercises, and solutions to all these exercises are given at the end of the module so that students can immediately check their understanding of each concept as it arises. Also, they have a model solution to follow if they encounter any difficulty. Standard sets of exercises follow each section and each unit, and answers to the odd-numbered exercises of these sets are given in the back of the module. Therefore, the material is self-paced and may be used in courses employing individualized instruction techniques, in mathematics laboratories, by students following a self-study program, or in more traditional courses.

The instructor's manual contains two unit tests for each unit of the module. These tests also may be used in individualized courses when pre-tests and post-tests for each unit are desired.

I wish to thank the members of the Cummings Publishing Company editorial staff who made valuable suggestions in the preparation of this module. Their assistance and cooperation are greatly appreciated.

A. J. W.

CONTENTS

UNIT ONE: ANGLES, TRIANGLES, AND QUADRILATERALS   1

1-1   Basic Angles 1
       Exercises 1-1-Section 5
1-2   Pairs of Angles 6
       Exercises 1-2-Section 9
1-3   Triangles 11
       Exercises 1-3-Section 14
1-4   Quadrilaterals 16
       Exercises 1-4-Section 19
       Exercises for Unit One 21

UNIT TWO: BASIC GEOMETRIC MEASURES AND PROPERTIES   24

2-1   Perimeter 24
       Exercises 2-1-Section 29
2-2   Area 32
       Exercises 2-2-Section 37
2-3   The Pythagorean Theorem 39
       Exercises 2-3-Section 42
2-4   Similar Triangles 43
       Exercises 2-4-Section 48
       Exercises for Unit Two 50

UNIT THREE: THE CIRCLE   54

3-1   Basic Definitions and Properties 54
       Exercises 3-1-Section 57
3-2   Circumference and Length of Arc 58
       Exercises 3-2-Section 62
3-3   Area 63
       Exercises 3-3-Section 66
       Exercises for Unit Three 67

UNIT FOUR: SOLID GEOMETRIC FIGURES   70

    4-1   Prisms 70
             Exercises 4-1-Section 74
    4-2   Cylinders 76
             Exercises 4-2-Section 79
    4-3   Pyramids and Cones 81
             Exercises 4-3-Section 85
    4-4   The Sphere 87
             Exercises 4-4-Section 90
             Exercises for Unit Four 90

Units of Measurement   94

Solutions for All Exercises of Short Exercise Sets and Answers for Odd-Numbered Exercises of Section and Unit Exercises   95

Appendix A: Significant Digits and Rounding Off   111

Appendix B: Table of Squares and Square Roots   113

Index   119

UNIT ONE: ANGLES, TRIANGLES, AND QUADRILATERALS

1-1 BASIC ANGLES

1-1-1 The analysis of a great many applied problems in science and technology requires the use of the topics and methods of geometry. Therefore, geometry along with other basic areas of mathematics such as arithmetic and algebra allows us to solve a great variety of problems encountered by a technician. It is our purpose in this module to introduce some of the basic concepts and methods of geometry.

1-1-2 Geometry deals with the properties and measurements of angles, lines, and surfaces and the basic figures which they form. In this unit we shall develop the terminology and basic properties associated with angles, triangles (three-sided figures), and quadrilaterals (four-sided figures).

1-1-3 It is not possible to define every word and prove every statement, and therefore certain words and concepts must be accepted without definition. Generally, in geometry, the concepts of a *point,* a *line,* and a *plane* are accepted as being known intuitively. We can then define other terms with the help of these undefined terms.

1-1-4 If two lines meet at point P, as shown in the figure, the amount of rotation to bring one line together with the other is called the *angle* through which the first was rotated. Notation and terminology associated with angles are illustrated in the following article.

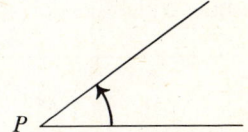

1-1-5 **EXAMPLE 1-1-A**

In the figure at the right, the angle formed by lines AB and AC is denoted as $\angle BAC$, or $\angle CAB$. The *vertex* of the angle is point A, and the *sides* of the angle are AB and AC.

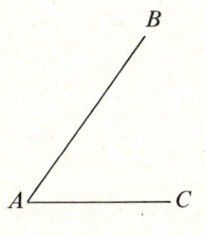

EXERCISES 1-1-A

Following Example 1-1-A, determine the following for the angle formed by lines OP and OQ.

1. Two designations of the angle
2. The vertex
3. The sides of the angle

1-1-6   One complete rotation of a line about a point is defined to be an angle of 360 *degrees*, written as 360°, as shown in the top figure at the right. A *straight angle* contains 180°, as shown in the bottom figure at the right. Another way of stating this would be to say that the *measures* of these angles are 360° and 180°, respectively.

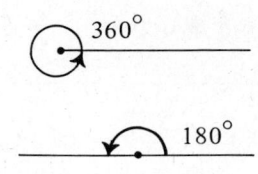

1-1-7   **EXAMPLE 1-1-B**

In the figure at the right, the angle around *A* is an angle of 360°. The angle shown at *B* is a straight angle, which means that it has a measure of 180°. The straight angle may be designated as ∠*ABC*, or ∠*CBA*.

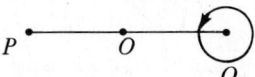

EXERCISES 1-1-B

In the figure below, determine the number of degrees in the given angles.

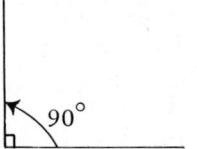

1. The angle around *Q*
2. ∠*POQ*

1-1-8   An angle whose measure is 90° is called a *right angle*. In a figure, a right angle is usually denoted by ⌐. If two lines meet so that the angle between them is 90°, the lines are said to be *perpendicular*. See the angle and lines shown in the figure to the right.

1-1-9   **EXAMPLE 1-1-C**

In the figure at the right, *AC* is perpendicular to *BD*. This is written as *AC* ⊥ *BD*. Therefore, ∠*ABD* and ∠*CBD* are right angles. This means that ∠*ABD* and ∠*CBD* have measures of 90°.

EXERCISES 1-1-C

In the figure, *PQ* is perpendicular to *OR*.
1. Write the statement for perpendicularity with proper notation.
2. In the figure, identify the right angles.

1-1-10   A device that can be used to measure an angle is a *protractor*. See the figure at the right. A protractor is used by placing the point in the middle of the bottom edge on the vertex of the angle, and the 0° line along one side. The number of degrees in the angle is found by noting where the other side of the angle crosses the edge of the protractor.

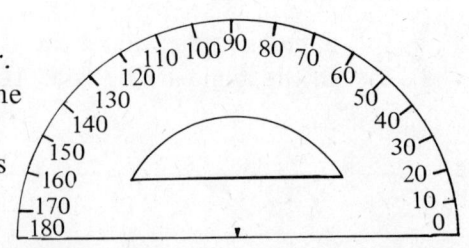

1-1-11 **EXAMPLE 1-1-D**

Measuring ∠ABC with a protractor, we place

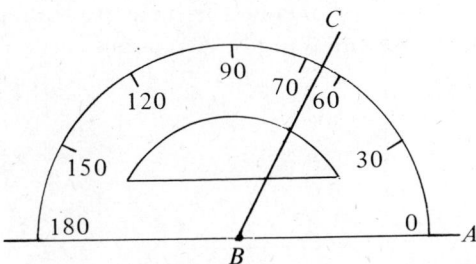

the point of the protractor on the vertex B and the 0° line along BA. Since BC crosses the 65° mark, we have the measure of ∠ABC as 65°. We will note this as ∠ABC = 65°.

EXERCISES 1-1-D

By use of a protractor, measure ∠ABC in each figure.

1.

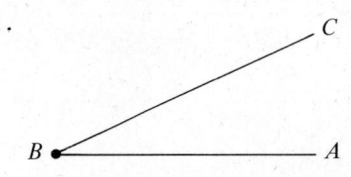

2.

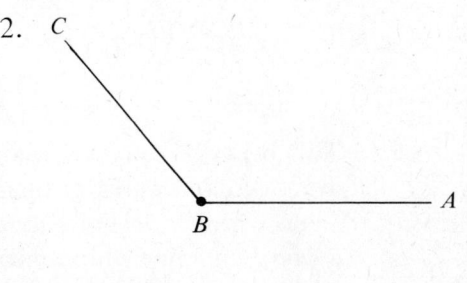

1-1-12 When it is necessary to measure angles to an accuracy of less than one degree, decimal parts of a degree may be used. This is becoming more common with the greater use of small calculators. However, previously the more common way to express such angles was to divide each degree into 60 equal parts, called *minutes*, and to divide each minute into 60 equal parts, called *seconds*. It is rather common for surveyors and astronomers to measure angles to the nearest second.

1-1-13 The notation for minutes is ′ and that for seconds is ″. Thus, an angle of 32 degrees, 15 minutes and 38 seconds is denoted as 32°15′38″. Since 1° = 60′, we find a decimal part of a degree in minutes by multiplying the decimal by 60′. Also, since $1' = (\frac{1}{60})°$, we express minutes in terms of a degree by dividing by 60. (Similar conversions can be used for seconds.)

4 Angles, Triangles, and Quadrilaterals

1-1-14 **EXAMPLE 1-1-E**

(a) $0.2° = (0.2)(60') = 12'$
Thus, $17.2° = 17°12'$.

(b) $36' = \left(\dfrac{36}{60}\right)° = 0.6°$.
Thus, $58°36' = 58.6°$.

(c) $6.35° = 6°21'$ since $(0.35)(60') = 21'$.

(d) $74°9' = 74.15°$ since $9' = \left(\dfrac{9}{60}\right)° = 0.15°$.

EXERCISES 1-1-E

In Exercises 1 and 2, change the measure to degrees and minutes. In Exercises 3 and 4, change the measure to degrees and decimal parts of a degree.

1. $41.3°$
2. $12.75°$
3. $86°42'$
4. $7°39'$

1-1-15 Two other basic types of angles are identified by whether they have measures which are less than or greater than 90°. An *acute angle* is an angle whose measure is between 0° and 90°. An *obtuse angle* is an angle whose measure is between 90° and 180°. These are shown in the figures at the right.

acute angle   obtuse angle

1-1-16 **EXAMPLE 1-1-F**

(a) An angle whose measure is 73° is an acute angle, since 73° is between 0° and 90°.

(b) An angle whose measure is 154° is an obtuse angle, since 154° is between 90° and 180°.

EXERCISES 1-1-F

The number of degrees in each of the following exercises is the measure of an angle. Identify the angles as acute or obtuse.

1. $83°$; $102°$; $90°30'$
2. $160°$; $89°59'$; $9°$

1-1-17 In this section we have discussed the meaning of an angle, and the measure of an angle in degrees, decimal parts of a degree, and minutes and seconds. A straight angle has a measure of 180°, and a right angle has a measure of 90°. Two lines which meet in a right angle are perpendicular. An acute angle has a measure between 0° and 90°, and an obtuse angle has a measure between 90° and 180°. The following exercises provide an opportunity to review the terminology and methods of this section.

1-1-18 EXERCISES 1-1-Section

In Exercises 1-4, use the figure at the right and identify the following:

1. Two designations of the angle with vertex at C
2. Two designations of the angle with vertex at B
3. The right angle
4. The perpendicular lines

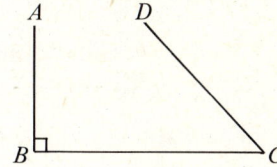

In Exercises 5-8, use the figure at the right and identify the following:

5. The sides of $\angle DBA$
6. The straight angle
7. The acute angle
8. The obtuse angle

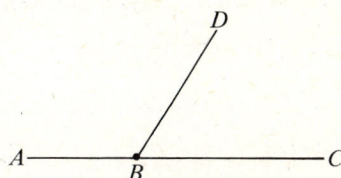

In Exercises 9-12, use the figure at the right and identify the following:

9. The perpendicular lines
10. The right angle
11. The obtuse angle
12. The acute angle

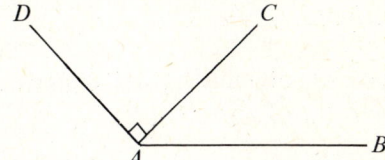

In Exercises 13-20, use the figure at the right and determine the measure of each of the indicated angles. Also, identify each as either acute or obtuse.

13. $\angle AOB$
14. $\angle AOC$
15. $\angle AOD$
16. $\angle AOE$
17. $\angle BOE$
18. $\angle BOF$
19. $\angle DOF$
20. $\angle EOF$

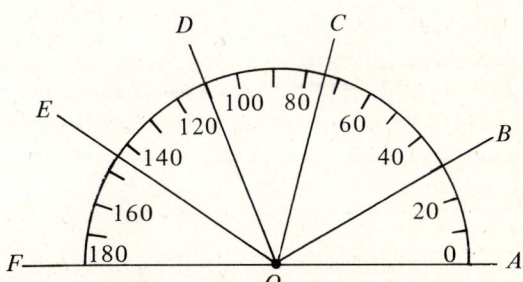

In Exercises 21-24, change the measure to degrees and minutes.

21. $56.4°$
22. $18.9°$
23. $136.45°$
24. $79.05°$

6   Angles, Triangles, and Quadrilaterals

In Exercises 25-28, change the measure to degrees and decimal parts of a degree.

25. $156°15'$ 26. $33°48'$ 27. $67°6'$ 28. $16°57'$

In Exercises 29-30, draw an appropriate figure.

29. A jet takes off from the horizontal ground at an angle of $23°$. Draw an appropriate angle and label its sides.

30. The foot of a ladder is on the horizontal ground, and it leans against a wall which is perpendicular to the ground. The angle between the ground and ladder is $65°$. Draw an appropriate figure, labeling the parts in the figure.

1-2   **PAIRS OF ANGLES**

1-2-1   When discussing certain geometric concepts we will find it necessary to refer to various pairs of angles. In some cases it will be angles of a particular type, and in other cases it will be the sum of the measures of two angles. The basic types of pairs of angles are discussed in this section.

1-2-2   Two angles which have a common vertex and a side common between them are known as *adjacent angles*.

1-2-3   **EXAMPLE 1-2-A**

In the figure at the right, $\angle BAC$ and $\angle CAD$ have a common vertex at A, and they have the common side AC between them. Therefore, $\angle BAC$ and $\angle CAD$ are adjacent angles.

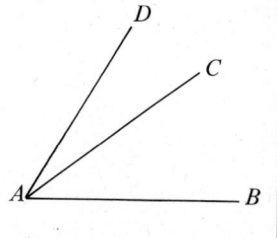

EXERCISES 1-2-A

1. Identify the adjacent angles in the figure at the right.

2. Draw a figure with adjacent angles $\angle RST$ and $\angle TSU$.

1-2-4   Two lines that cross form equal angles on opposite sides of the point of intersection, which is the common vertex. These equal angles are called *vertical angles*.

1-2 Pairs of Angles

1-2-5   **EXAMPLE 1-2-B**

In the figure at the right, lines AB and CD intersect at O. $\angle AOC$ and $\angle BOD$ are vertical angles, and they are equal. Also, $\angle BOC$ and $\angle AOD$ are vertical angles and are equal. Thus, we may state that vertical angles are equal.

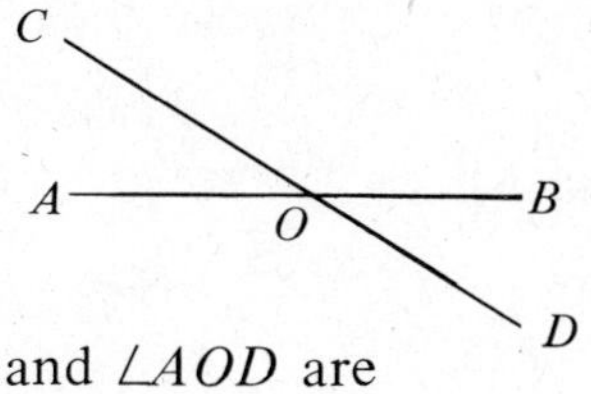

EXERCISES 1-2-B

1. Identify the two pairs of vertical angles in the figure at the right. 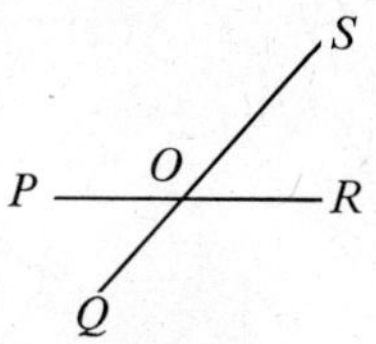
2. Lines DE and FG intersect at point H. Draw the figure and identify the pairs of vertical angles.

1-2-6   If the sum of the measures of two angles is 180°, the angles are called *supplementary angles*.

1-2-7   **EXAMPLE 1-2-C**

(a) In the figures at the right, $\angle BAC = 55°$ and $\angle DEF = 125°$. Since $\angle BAC + \angle DEF = 55° + 125° = 180°$, $\angle BAC$ and $\angle DEF$ are supplementary angles.

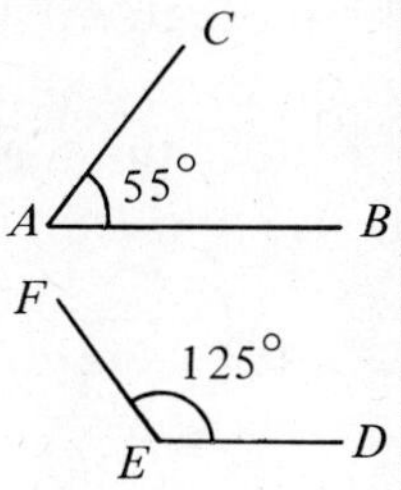

(b) In the figure at the right, $\angle POQ$ is a straight angle, which means $\angle POQ = 180°$. Since $\angle QOR + \angle POR = \angle POQ = 180°$, $\angle QOR$ and $\angle POR$ are supplementary angles.

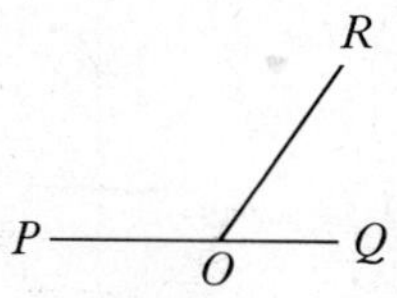

EXERCISES 1-2-C

1. $\angle RST = 70°$, $\angle UVW = 20°$, and $\angle XYZ = 110°$. Which of these angles are supplementary?
2. Draw a figure with straight angle LMN and line PM. Identify the supplementary angles.

1-2-8   If the sum of the measures of two angles is 90°, the angles are called *complementary angles*.

8   Angles, Triangles, and Quadrilaterals

1-2-9 **EXAMPLE 1-2-D**

(a) In the figures at the right, ∠BAC = 55° and ∠GHJ = 35°. Since ∠BAC + ∠GHJ = 55° + 35° = 90°, ∠BAC and ∠GHJ are complementary angles.

(b) In the figure at the right, ∠STU is a right angle, which means ∠STU = 90°. Since ∠STV + ∠VTU = ∠STU = 90°, ∠STV and ∠VTU are complementary angles.

EXERCISES 1-2-D

1. ∠RST = 70°, ∠UVW = 20°, and ∠XYZ = 110°. Which of these angles are complementary?

2. Draw a figure with right angle EFG and line HF such that ∠EFH is acute. Identify the complementary angles.

1-2-10   Since the sum of the measures of supplementary angles is 180°, if the measure of one angle is known, the measure of the other, its *supplement*, can be found by subtracting the known measure from 180°. Also, since the sum of the measures of complementary angles is 90°, the measure of one can be found by subtracting the measure of the other, its *complement*, from 90°.

1-2-11 **EXAMPLE 1-2-E**

(a) The measure of an angle is 36°. The measure of its supplement is 144°, since

    180° − 36° = 144°.

(b) The measure of an angle is 36°. The measure of its complement is 54°, since

    90° − 36° = 54°.

EXERCISES 1-2-E

1. Determine the measure of the supplement of an angle whose measure is 59°.

2. Determine the measure of the complement of an angle whose measure is 59°.

3. In the figure at the right, identify and find the measure of the supplement of ∠ABC.

4. In the same figure, identify and find the measure of the complement of ∠ABC.

1-2-12   In a plane, if a line crosses two *parallel* or nonparallel lines, it is called a *transversal*. Parallel lines are lines whose extensions will not meet. In the figure at the right, the transversal EF crosses a pair of parallel lines. We denote the fact that AB is parallel to CD by AB ∥ CD.

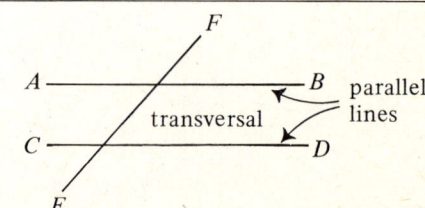

1-2-13 When a transversal crosses a pair of parallel lines, certain pairs of angles with equal measures result. In the figure at the right, the measures of the *corresponding angles* are equal. (That is, ∠1 = ∠5, ∠2 = ∠6, ∠3 = ∠7, and ∠4 = ∠8). Also, the measures of the *alternate-interior angles* are equal (∠3 = ∠6 and ∠4 = ∠5). [The measures of the *alternate-exterior* angles are also equal (∠1 = ∠8 and ∠2 = ∠7).] Note the use of numbers to conveniently label the angles.

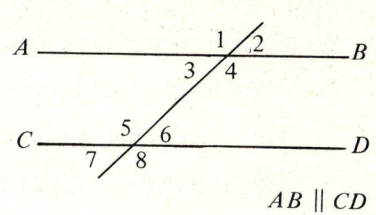

AB ∥ CD

1-2-14 **EXAMPLE 1-2-F**

In the figure at the right, ∠1 = ∠5 since these angles are corresponding angles. Also, ∠4 = ∠5 since these angles are alternate-interior angles.

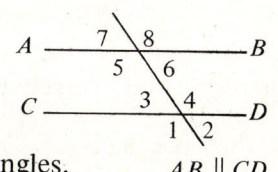

AB ∥ CD

EXERCISES 1-2-F

In the figure with Example 1-2-F, identify another pair of

1. corresponding angles,
2. alternate-interior angles.

1-2-15 In this section we have discussed several important pairs of angles. These pairs of angles include adjacent angles, vertical angles, supplementary angles, complementary angles, and corresponding angles. Also, with parallel lines and a transversal we discussed corresponding angles and alternate-interior angles. The following exercises provide an opportunity to review these pairs of angles.

1-2-16 **EXERCISES 1-2-Section**

In Exercises 1-2, use the figure at the right. Identify the indicated pairs of angles.

1. Two pairs of vertical angles
2. Two pairs of supplementary angles

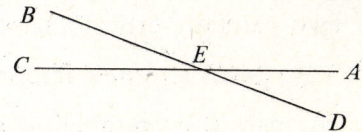

In Exercises 3-4, use the figure at the right. Identify the indicated pairs of angles.

3. Two pairs of adjacent angles
4. One pair of complementary angles

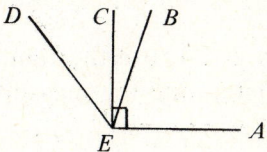

10 Angles, Triangles, and Quadrilaterals

In Exercises 5-8, use the figure at the right. Identify the indicated angles.

5. One pair of supplementary right angles
6. Two pairs of adjacent angles
7. The complement of ∠DBE
8. The supplement of ∠CBD

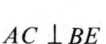
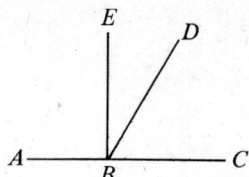

AC ⊥ BE

In Exercises 9-12, use the figure at the right. Determine the measures of the indicated angles.

9. ∠DBE
10. ∠EBF
11. ∠DBA
12. ∠FBA

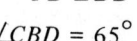
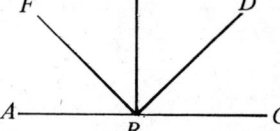

AC ⊥ BE
FB ⊥ DB
∠CBD = 65°

In Exercises 13-14, use the figure at the right. Determine the measures of the indicated angles.

13. ∠4
14. ∠3

∠1 = ∠2

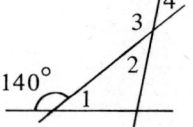

In Exercises 15-16, use the figure at the right. Determine the measures of the indicated angles.

15. ∠EOF
16. ∠EOC

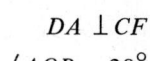
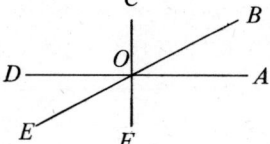

DA ⊥ CF
∠AOB = 28°

In Exercises 17-20, use the figure at the right. Identify the indicated pairs of angles from those which are numbered.

17. Two pairs of vertical angles
18. One pair of alternate-interior angles
19. One pair of corresponding angles
20. One pair of supplementary angles

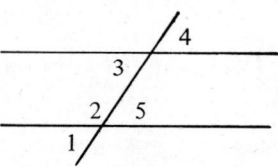

In Exercises 21-24, use the figure at the right (lines which appear parallel are parallel). Determine the measures of the indicated angles.

21. ∠1
22. ∠2
23. ∠4
24. ∠3

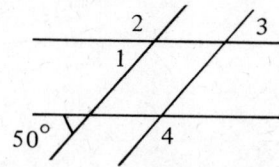

In Exercises 25-28, use the figure at the right. Determine the measures of the indicated angles.

25. ∠FCE   26. ∠ECD

27. ∠BCE   28. ∠BFC

BF ∥ CE
FC ⊥ AD
∠ABF = 148°

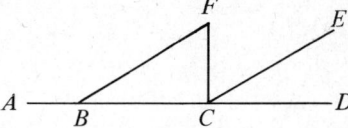

1-3   TRIANGLES

1-3-1   When a part of the plane is bounded and closed by straight line segments, it is called a *polygon*. In general, polygons are named according to the number of sides they have. A *triangle* has three sides, a *quadrilateral* has four sides, a *pentagon* has five sides, a *hexagon* has six sides, and so on. The polygons of greatest general importance are the triangle and the quadrilateral. Therefore, this section is devoted to the triangle and the following section is devoted to the quadrilateral.

1-3-2   A triangle has three sides, and these three sides form three interior angles. In an *equilateral triangle* the three sides are equal in length, and the measure of each of the angles is 60°. In an *isosceles triangle* two of the sides are equal in length, and the measures of the *base angles* (the angles opposite the equal sides) are also equal. In a *scalene triangle* no two sides are equal in length. At the right, the upper triangle is equilateral since all sides are of length s, the middle triangle is isosceles since two sides are of length s, and the bottom triangle is scalene.

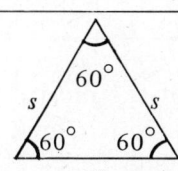
equilateral triangle

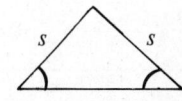
isosceles triangle

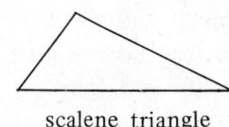
scalene triangle

12 Angles, Triangles, and Quadrilaterals

1-3-3 **EXAMPLE 1-3-A**

(a) The triangle at the right, denoted as △ABC, is equilateral. Since side AC = 5, we also know that AB = 5 and BC = 5. The angles also have a measure of 60°. We note this by writing ∠A = 60°, ∠B = 60°, and ∠C = 60°.

(b) The triangle at the right is isosceles since DF = 7 and EF = 7. The base angles are ∠D and ∠E. Therefore, since ∠D = 70°, we know that ∠E = 70°.

(c) The triangle at the right, △PQR, is scalene since none of the sides is equal in length to another of the sides.

EXERCISES 1-3-A

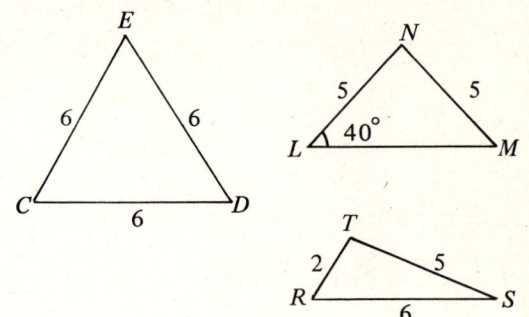

1. Identify the isosceles triangle.
2. Identify the scalene triangle.
3. Identify the equilateral triangle.
4. In △CDE, what is the measure of ∠D?
5. In the isosceles triangle, identify the base angles.
6. In △LMN, what is the measure of ∠M?

1-3-4 One of the most important types of triangles in scientific and technical applications is the *right triangle*. In a right triangle, one of the angles is a right angle. The side opposite the right angle is called the *hypotenuse*, and the other two sides are called the *legs*. The triangle at the right is a right triangle.

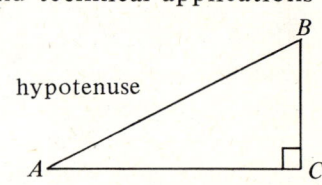

1-3-5 **EXAMPLE 1-3-B**

The triangle at the right, △PQR, is a right triangle since ∠Q = 90°. The hypotenuse is side PR, and the legs are PQ and QR.

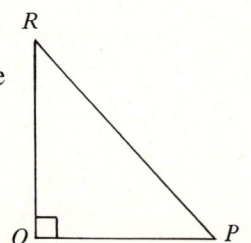

EXERCISES 1-3-B

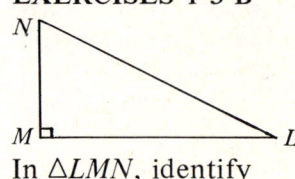

In △LMN, identify
1. the hypotenuse,
2. the legs.

1-3-6 One important property of a triangle is that the sum of the measures of its angles is 180°. We can prove this as follows.
In the figure at the right $EC \parallel AB$. Since $\angle 1$, $\angle 2$, and $\angle 3$ constitute a straight angle,

$$\angle 1 + \angle 2 + \angle 3 = 180°.$$

Also, $\angle 1 = \angle 4$ and $\angle 3 = \angle 5$ since each pair is a pair of alternate-interior angles. Thus,

$$\angle 4 + \angle 2 + \angle 5 = 180°.$$

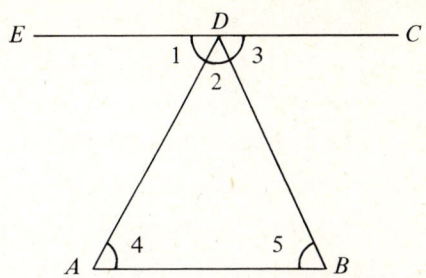

1-3-7 Therefore, we conclude that *the sum of the measures of the angles of a triangle is 180°*. Thus, if the measures of two angles of a triangle are known, the measure of the third may be determined by subtracting the sum of the first two from 180°.

1-3-8 **EXAMPLE 1-3-C**

(a) In the triangle at the right, we may find the measure of $\angle A$ as follows:

$$\angle B + \angle C = 55° + 80° = 135°$$
$$\angle A = 180° - 135° = 45°$$

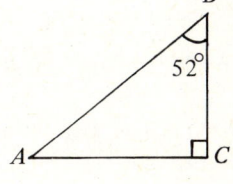

(b) In the triangle at the right, $\angle C = 90°$ since it is denoted as a right angle. Therefore, we may determine the measure of $\angle A$ as follows:

$$\angle B + \angle C = 52° + 90° = 142°$$
$$\angle A = 180° - 142° = 38°$$

EXERCISES 1-3-C

In each of the indicated triangles, determine the measure of $\angle A$.

1.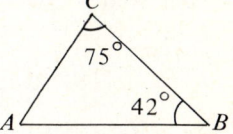

2. In $\triangle ABC$, $\angle B = 72°$ and $\angle C = 69°$.

3.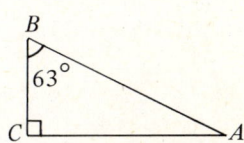

4. $\triangle ABC$ is a right $\triangle$, with right angle at C and $\angle B = 17°$.

14 Angles, Triangles, and Quadrilaterals

1-3-9   **EXAMPLE 1-3-D**

In △ABC at the right, we find the measure of ∠ACB as follows:

∠A = 38° since it and ∠DCA are alternate-interior angles.

∠B = 90° since it is a right angle.

∠A + ∠B = 38° + 90° = 128°

∠ACB = 180° − 128° = 52°

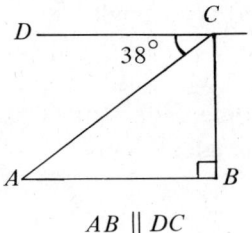

AB ∥ DC

EXERCISES 1-3-D

In each triangle determine the measure of ∠A.

1.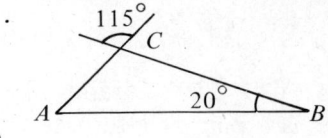

2.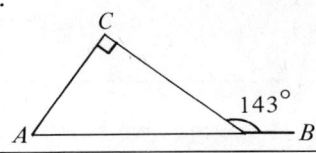

1-3-10 In this section we have introduced the basic types of triangles. These include the equilateral triangle, isosceles triangle, scalene triangle, and right triangle. We also noted that the sum of the measures of the angles of a triangle is 180°. The following exercises provide an opportunity to review the material of this section.

1-3-11   **EXERCISES 1-3-Section**

In Exercises 1-4, use the figure at the right. In the figure note △ADC, △ADB, and △DBC.

1. Identify the isosceles triangle.
2. Identify the scalene triangle.
3. Identify the right triangle.
4. ∠C = 53°, determine the measure of ∠DBA.

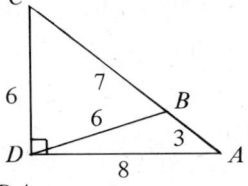

In Exercises 5-8, use the figure at the right. In the figure △DEC is equilateral, EC ∥ AB, EC = 5, and AE = 2.

5. Determine the length DC.
6. Determine the length AD.
7. Determine the measure of ∠AEC.
8. Determine the measure of ∠B.

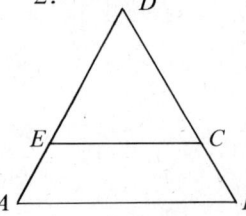

In Exercises 9-16, find the measure of ∠A.

9. 10. 11. 12.

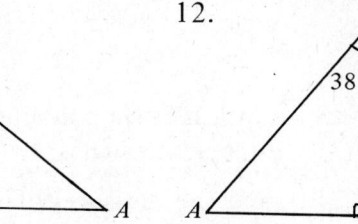

13. 14. 15. 16.

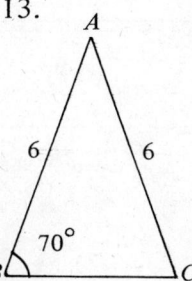 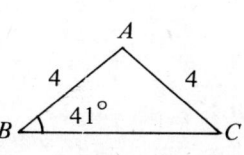 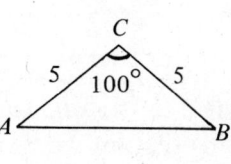 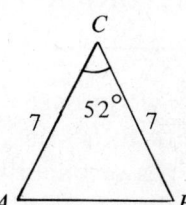

In Exercises 17-20, determine the measure of the indicated angle.

17. ∠CBD 18. ∠DBC 19. ∠ACB 20. ∠C

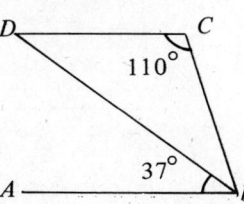

AB ∥ DC

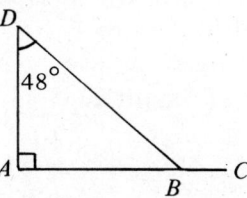

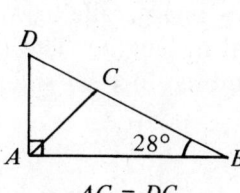

AC = DC

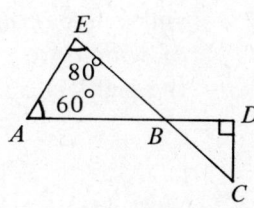

In Exercises 21-24, solve the given problems.

21. The streets in a certain city meet at the angles shown in the figure at the right. Find the indicated angle (labeled x).

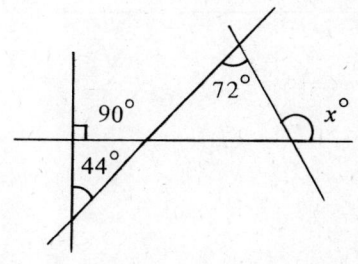

22. Metal braces support a beam as shown in the figure at the right. Find the indicated angle.

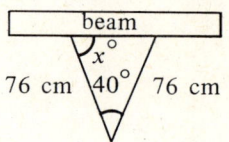

23. Can a triangle contain two right angles? an obtuse angle and a right angle? three acute angles?

24. Find the measures of the angles of an isosceles right triangle (the triangle is isosceles and contains a right angle).

1-4 QUADRILATERALS

1-4-1   As we noted in Article 1-3-1, a *quadrilateral* is a polygon with four sides. In this section we discuss the important types of quadrilaterals and some of their properties. The figure at the right shows a general quadrilateral.

1-4-2   A *parallelogram* is a quadrilateral in which opposite sides are equal in length, and opposite angles have equal measures. The upper figure at the right is a parallelogram since opposite sides are of lengths a and b. A *rhombus* is a parallelogram, all four sides of which are equal in length. The lower figure at the right is a rhombus since all sides are of length a.

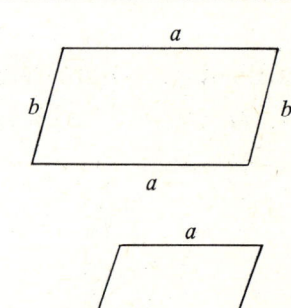

1-4 Quadrilaterals

1-4-3 EXAMPLE 1-4-A

(a) In parallelogram ABCD, opposite sides are equal. This means that $AB = DC$ and $AD = BC$. Also, opposite angles are equal in measure. This means that $\angle A = \angle C$ and $\angle B = \angle D$.

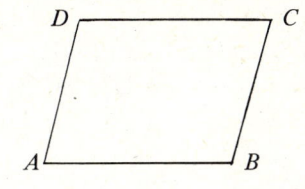

(b) In rhombus EFGH, all sides are equal. Therefore, $EF = FG = GH = HE$. Also, since a rhombus has the properties of a parallelogram, opposite angles have equal measures. Therefore, $\angle E = \angle G$ and $\angle F = \angle H$.

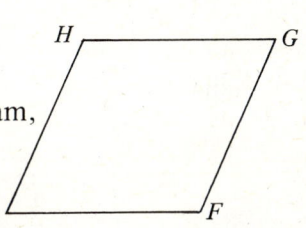

EXERCISES 1-4-A

In parallelogram ABCD, determine the indicated quantities.

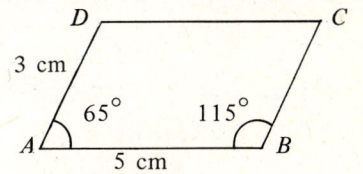

1. length of BC
2. length of CD
3. measure of $\angle C$
4. measure of $\angle D$

In rhombus EFGH, determine the indicated quantities.

5. length of FG
6. length of EH
7. measure of $\angle G$
8. measure of $\angle H$

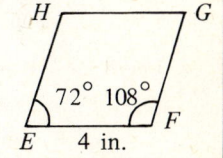

1-4-4 A *rectangle* is a parallelogram in which intersecting sides are perpendicular, which means that all four interior angles are right angles. Also, opposite sides of a rectangle are equal in length and parallel. The upper figure at the right is a rectangle. A *square* is a rectangle all four sides of which are equal. The lower figure at the right is a square.

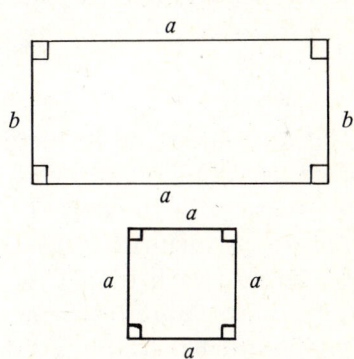

1-4-5 EXAMPLE 1-4-B

(a) In rectangle ABCD, the opposite sides are equal. This means that $AB = CD$ and $BC = AD$. All interior angles are right angles. This means that $\angle A = \angle B = \angle C = \angle D = 90°$.

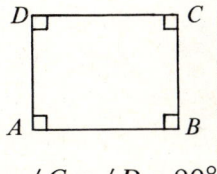

(b) In square EFGH, all four sides are equal and all four interior angles are right angles. This means that $EF = FG = GH = HE$ and $\angle E = \angle F = \angle G = \angle H = 90°$.

EXERCISES 1-4-B

In rectangle ABCD, determine the indicated quantities.

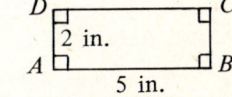

1. length of BC
2. length of CD

In square EFGH, determine the indicated quantities.

3. length of FG
4. measure of $\angle H + \angle E$

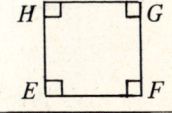

1-4-6 A *trapezoid* is a quadrilateral in which two of the sides are parallel. These parallel sides are called the *bases* of the trapezoid. The quadrilateral at the right is a trapezoid where the bases are labeled b_1 and b_2.

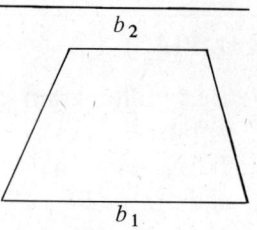

1-4-7 **EXAMPLE 1-4-C**

In quadrilateral $ABCD$, $AB \parallel CD$. Therefore, it is a trapezoid with bases AB and CD. The *base angles* of AB are $\angle A$ and $\angle B$, and the base angles of CD are $\angle C$ and $\angle D$.

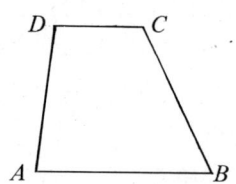

EXERCISES 1-4-C

In trapezoid $EFGH$, $EF \parallel HG$.

1. Identify the bases.
2. Identify the base angles.

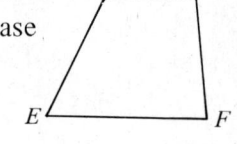

1-4-8 When we are identifying quadrilaterals, more than one designation may be strictly correct. For example, a square can also be considered to be a rectangle (four right angles) or a rhombus (four equal sides). However, only a square has both properties, and the word "square" should be used to identify the figure. The word which is most descriptive should be used in all cases.

1-4-9 Consider a quadrilateral to be divided into two triangles as shown in the figure at the right. (The line drawn from one vertex to the opposite vertex is called a *diagonal* of the quadrilateral.) The sum of the measures of the angles of each triangle is 180°. Since these angles give the total measure of the angles of the quadrilateral, we conclude that *the sum of the measures of the angles of a quadrilateral is 360°*.

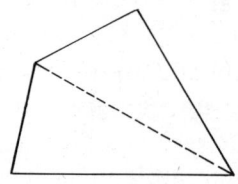

1-4-10 **EXAMPLE 1-4-D**

(a) In quadrilateral $ABCD$ at the right, we may find the measure of $\angle D$ as follows:

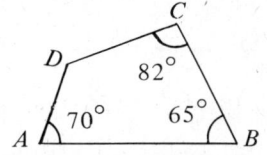

$\angle A + \angle B + \angle C = 70° + 65° + 82° = 217°$
$\angle D = 360° - 217° = 143°$

(b) In parallelogram $ABCD$ at the right, we find that the measure of $\angle C$ is the same as that for $\angle A$ since they are opposite angles. Thus, $\angle C = 68°$. Since $\angle B = \angle D$ (opposite angles), we see that twice the measure of $\angle B$ is the sum of measures of $\angle A$ and $\angle C$ subtracted from 360°. Thus,

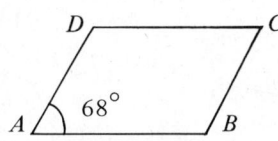

$\angle A + \angle C = 68° + 68° = 136°$
$2(\angle B) = 360° - 136° = 224°$
$\angle B = 112°$.

Thus, $\angle C = 68°$, $\angle B = 112°$, and $\angle D = 112°$.

EXERCISES 1-4-D

In quadrilateral $ABCD$, find the measure of $\angle A$.

1.   2.

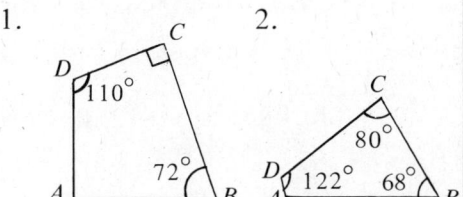

3. In parallelogram $ABCD$, find the measures of $\angle B$, $\angle C$, and $\angle D$.

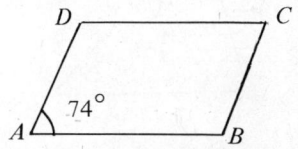

4. In rhombus $ABCD$, find the measures of $\angle B$, $\angle C$, and $\angle D$.

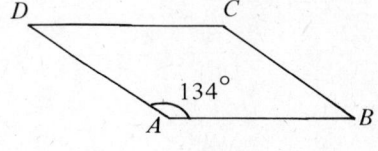

1-4-11 In this section we have introduced the basic types of quadrilaterals. These include the parallelogram, rhombus, rectangle, square, and trapezoid. We also noted that the sum of the measures of the angles of a quadrilateral is 360°. The following exercises provide an opportunity to review the material of this section.

1-4-12 **EXERCISES 1-4-Section**

In Exercises 1-8, draw (approximately) appropriate figures subject to the given conditions.

1. A quadrilateral with no two equal sides

2. A quadrilateral, with two opposite right angles, which is not a rectangle

3. A parallelogram with sides of 2 cm and 3 cm and an angle of 60° between them

20   Angles, Triangles, and Quadrilaterals

4. A rhombus with sides of 2 in. and an interior angle of 30°
5. A square with sides of 3 in.
6. A rectangle with sides of 2 cm and 4 cm
7. A trapezoid with bases 5 cm and 3 cm and unequal nonparallel sides
8. A trapezoid with bases 4 in. and 3 in. and equal nonparallel sides of 2 in.

In Exercises 9-12, find the measure of ∠A in quadrilateral ABCD.

9.   10.   11.   12.

In Exercises 13-16, find the measures of ∠B, ∠C, and ∠D in parallelogram ABCD.

13.   14.   15.   16.

If the nonparallel sides of a trapezoid are equal in length, the figure is an *isosceles trapezoid*. In an isosceles trapezoid the base angles at either base are equal in measure. In Exercises 17-20, find the measures of ∠B, ∠C, and ∠D in isosceles trapezoid ABCD.

17.   18.   19.   20.

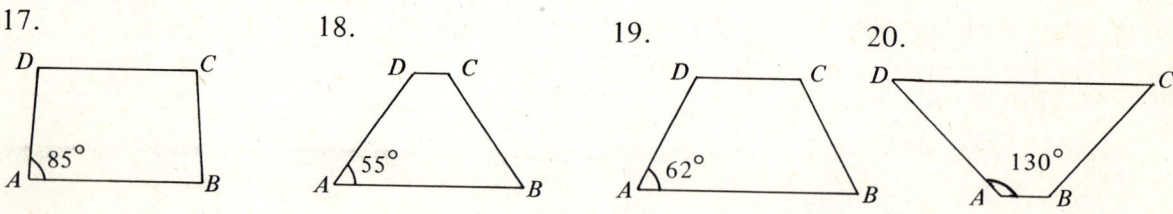

In Exercises 21-24, determine the measure of each of the angles of parallelogram ABCD.

21.

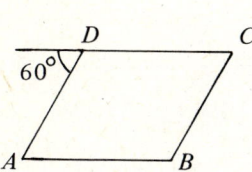

22.

23.

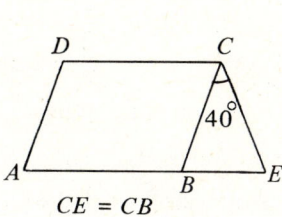

CE = CB

24.

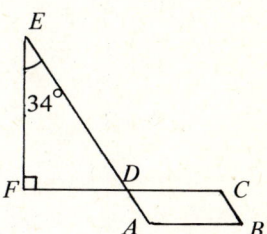

In Exercises 25-28, solve the given problem.

25. Show that the measures of the angles of a rectangle satisfy the statement in italics in Article 1-4-9.

26. Can the opposite sides of a quadrilateral be equal in length and the quadrilateral not be a parallelogram?

27. Draw a rhombus and its two diagonals. What conclusions can be drawn about the diagonals?

28. Draw a rectangle and one of its two diagonals. Into what figures is the rectangle divided?

1-5 EXERCISES FOR UNIT ONE

In Exercises 1-8, use the figure at the right and identify the following:

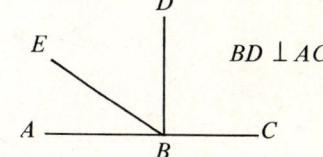

1. Two right angles
2. A straight angle
3. An obtuse angle
4. Two acute angles
5. The complement of ∠ABE
6. The supplement of ∠ABE
7. A pair of adjacent angles, one of which is a right angle
8. A pair of adjacent angles, neither of which is a right angle

In Exercises 9-12, use the figure at the right and identify the following (use only labeled angles):

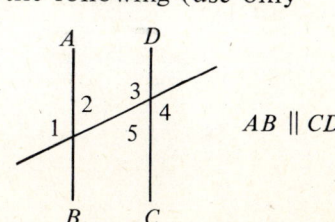

9. A pair of alternate-interior angles
10. A pair of corresponding angles
11. A pair of vertical angles
12. The measure of ∠1 + ∠5

22 Angles, Triangles, and Quadrilaterals

In Exercises 13-16, use the figure at the right. From among triangles *ABD*, *BCD*, and *ACD* and quadrilaterals *ABDE* and *ACDE* identify the following:

13. Two right triangles

14. An isosceles triangle

15. A rectangle

16. A trapezoid

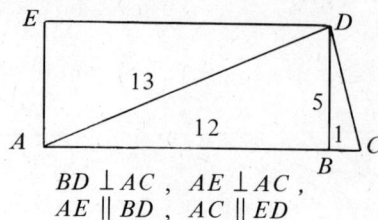

$BD \perp AC$, $AE \perp AC$,
$AE \parallel BD$, $AC \parallel ED$

In Exercises 17-18, change the measure to degrees and minutes. In Exercises 19-20, change the measure to degrees and decimal parts of a degree.

17. 43.95°   18. 12.55°   19. 105°54′   20. 215°45′

In Exercises 21-36, use the given figures. Determine the measures of the indicated angles.

21. ∠EBD
22. ∠EBC

23. ∠CBD
24. ∠ABC

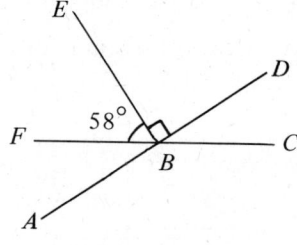

(Exer. 21–22)   (Exer. 23–24)

25. ∠BDC
26. ∠CDE

27. ∠BFD
28. ∠FDE

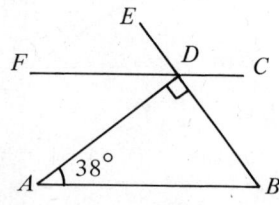

$AB \parallel FC$

(Exer. 25–26)

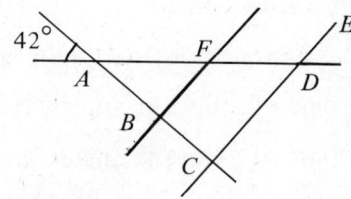

$BF \perp AC$
$EC \perp AC$

(Exer. 27–28)

29. ∠C 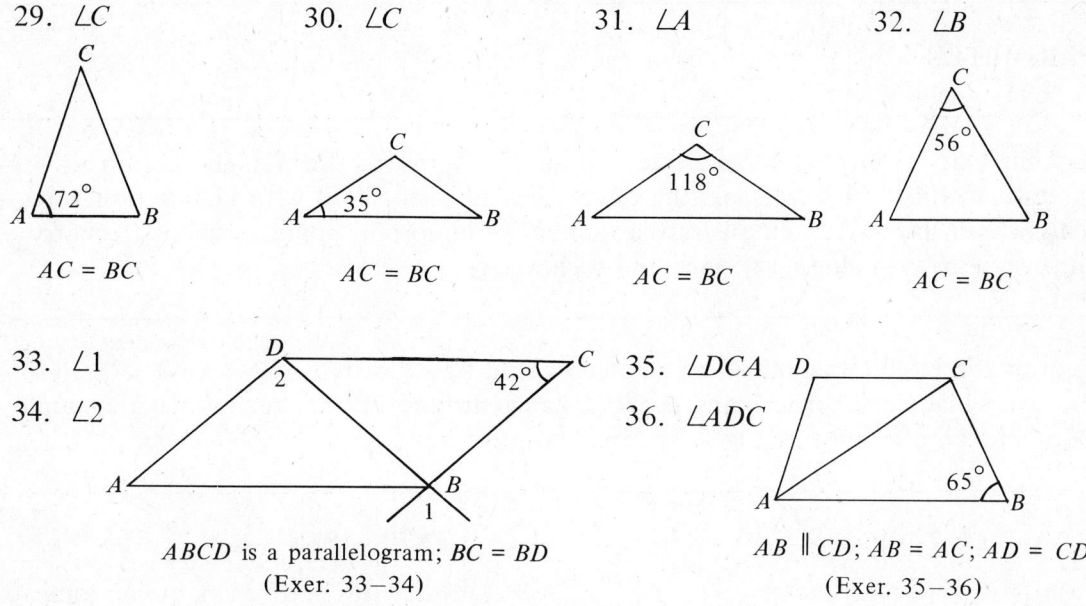 30. ∠C   31. ∠A   32. ∠B

AC = BC   AC = BC   AC = BC   AC = BC

33. ∠1
34. ∠2

ABCD is a parallelogram; BC = BD
(Exer. 33–34)

35. ∠DCA
36. ∠ADC

AB ∥ CD; AB = AC; AD = CD
(Exer. 35–36)

In Exercises 37-44, solve the given problems.

37. Show that two lines perpendicular to the same line are parallel. (Use equal corresponding angles.)

38. Can a triangle have sides of 3 cm, 4 cm, and 8 cm?

39. In the figure at the right, △ABE is equilateral and BE ∥ CD. Show that △ACD is equilateral.

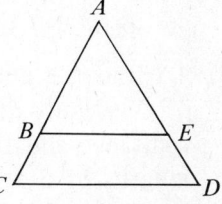

40. Draw a rhombus and one diagonal. Into what types of triangles is the rhombus divided?

41. Draw a right triangle with one leg twice as long as the other. Show that the *median* (line from vertex to midpoint of opposite side) to the hypotenuse is not the same as the *angle-bisector* (line that divides the angle in half) of the right angle.

42. Draw a pentagon (five-sided figure) and divide it into triangles by two diagonals. What conclusion is drawn about the sum of the measures of the interior angles?

43. When a ray of light strikes a plane mirror, the angle it makes with a line perpendicular to the mirror equals the angle between the reflected ray and the perpendicular line. If the angle between a light ray and a perpendicular to a mirror is 34°, what is the angle between the reflected ray and the mirror?

44. What must be the base angles of four identical isosceles trapezoid tables which when placed next to each other form a square table (with a square hole in the center)?

UNIT TWO: BASIC GEOMETRIC MEASURES AND PROPERTIES

2-1 PERIMETER

2-1-1 In Unit One we introduced the basic geometric figures of the triangle and quadrilateral. In this unit we discuss the basic measures associated with geometric figures. This will enable us to demonstrate a number of important applications of geometry in several areas, including science and technology.

2-1-2 One of the basic measures of a plane geometric figure is its *perimeter*. The perimeter of a plane geometric figure is the distance around it. In the following example, perimeters are found directly from this definition.

2-1-3 EXAMPLE 2-1-A

(a) To find the perimeter of the triangle shown at the right, we add the lengths of the sides. Thus, the perimeter p is

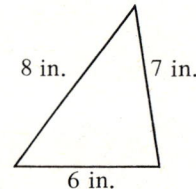

$p = 6$ in. $+ 7$ in. $+ 8$ in. $= 21$ in.

(b) To find the perimeter of the figure at the right, we add the lengths of the sides, even though the figure may appear to be more complicated. Therefore,

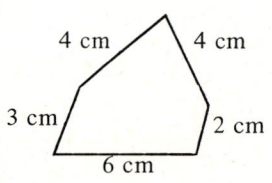

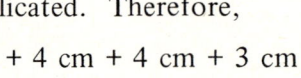
= 19 cm.

EXERCISES 2-1-A

Find the perimeters of the indicated figures.

1.   2.

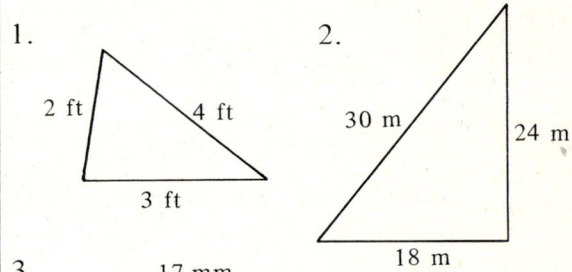

3.

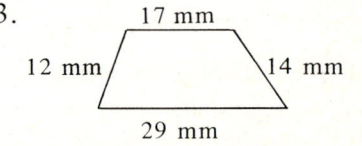

4.

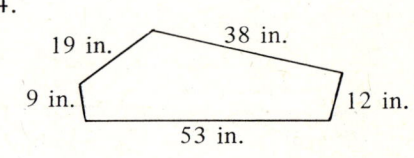

2-1-4 Using the definition of perimeter, we can derive formulas for the perimeters of many plane figures. However, if we remember the definitions of the figures and that of perimeter, we do not have to memorize most of these perimeter formulas.

2-1-5 The perimeter of a triangle with sides of lengths a, b, and c is the sum of a, b, and c. Thus,

$$p = a + b + c.$$

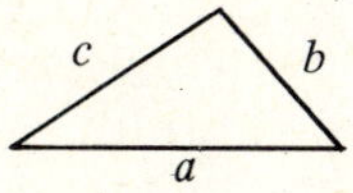

The perimeter of a quadrilateral with sides of lengths a, b, c, and d is the sum of a, b, c, and d. Thus,

$$p = a + b + c + d.$$

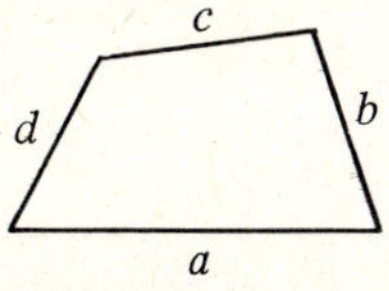

2-1-6

EXAMPLE 2-1-B

(a) The perimeter of a scalene triangle in which the lengths of the sides are $a = 15$ cm, $b = 18$ cm, and $c = 23$ cm is

$p = 15$ cm $+ 18$ cm $+ 23$ cm $= 56$ cm.

(b) The perimeter of a quadrilateral in which the lengths of the sides are $a = 29$ ft, $b = 14$ ft, $c = 28$ ft, and $d = 37$ ft is

$p = 29$ ft $+ 14$ ft $+ 28$ ft $+ 37$ ft $= 108$ ft.

EXERCISES 2-1-B

Find the perimeter of a triangle in which the lengths of the sides are as follows:

1. $a = 35$ in., $b = 41$ in., $c = 52$ in.
2. $a = 521$ mm, $b = 376$ mm, $c = 296$ mm

Find the perimeter of a quadrilateral in which the lengths of the sides are as follows:

3. $a = 79$ yd, $b = 63$ yd, $c = 46$ yd, $d = 91$ yd
4. $a = 317$ m, $b = 142$ m, $c = 256$ m, $d = 407$ m

2-1-7   Since the three sides of an equilateral triangle are equal in length, the perimeter is three times this length. Thus,

$$p = 3s.$$

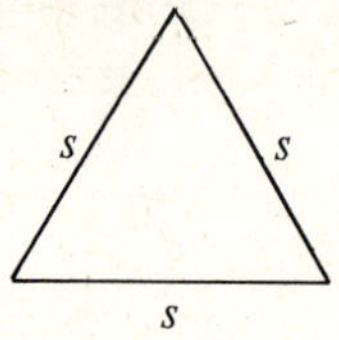

Since two sides of an isosceles triangle are equal in length, the perimeter is twice this length plus the length of the third side. Thus,

$$p = 2s + b$$

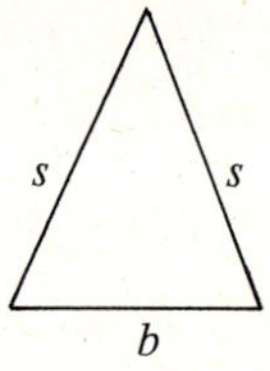

2-1-8   **EXAMPLE 2-1-C**

(a) The perimeter of an equilateral triangle of side 18 ft is

$p = 3(18 \text{ ft}) = 54 \text{ ft}.$

(b) The perimeter of an isosceles triangle with equal sides 128 cm and third side 56 cm is

$p = 2(128 \text{ cm}) + 56 \text{ cm} = 312 \text{ cm}.$

EXERCISES 2-1-C

Find the perimeters of an equilateral triangle in which the length of a side is:

1. 14 m   2. 520 yd

Find the perimeter of an isosceles triangle in which the lengths of the equal sides and third side are:

3. $s = 26$ cm, $b = 45$ cm

4. $s = 37$ in., $b = 19$ in.

2-1-9   Since the four sides of a square or rhombus are equal in length, the perimeter of each is four times the length of one of the sides. Thus,

$p = 4s.$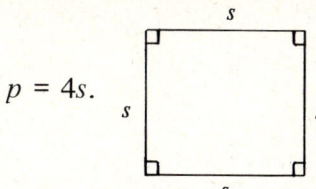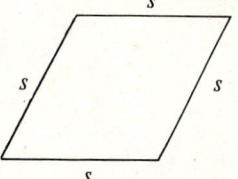

Since the opposite sides of a parallelogram or rectangle are equal in length, the perimeter of each is twice one of these lengths plus twice the other length. Thus,

$p = 2a + 2b.$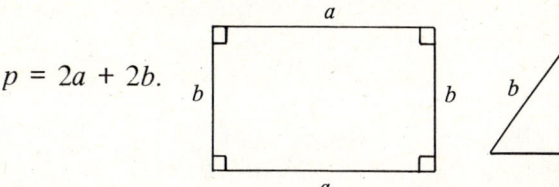

In a rectangle, the longer side is usually called the *length* and the shorter side is called the *width*.

2-1-10 **EXAMPLE 2-1-D**

(a) The perimeter of a square (or a rhombus) of side 63 cm is

$p = 4(63 \text{ cm}) = 252 \text{ cm}.$

(b) The perimeter of a rectangle of length 18 in. and width 14 in. is

$p = 2(18 \text{ in.}) + 2(14 \text{ in.})$
$ = 36 \text{ in.} + 28 \text{ in.} = 64 \text{ in.}$

EXERCISES 2-1-D

Find the perimeters of the indicated figures.

1. square: $s = 17$ ft
2. rhombus: $s = 46$ mm
3. rectangle: length = 29 cm
   width = 17 cm
4. parallelogram: $a = 15$ in., $b = 9$ in.

2-1-11 By using the definition of perimeter, we can find the perimeters of geometric figures that are combinations of basic figures. This is illustrated in the following article.

2-1-12 **EXAMPLE 2-1-E**

The figure at the right is a combination of a rectangle and an equilateral triangle. (The dashed line shows where the figures are joined, but it should not be counted as part of the perimeter since it is not on the outside of the figure). Since the lower part of the figure is a rectangle, we see that the dashed line is 32 cm long. This allows us to see that each of the sides of the triangle is 32 cm long, since it is equilateral. Thus, the perimeter is found by adding the 32 cm along the bottom to two 32-cm lengths at the top to two 45-cm lengths at the sides. Thus,

p = 32 cm + 2(32 cm) + 2(45 cm)
   = 32 cm + 64 cm + 90 cm = 186 cm.

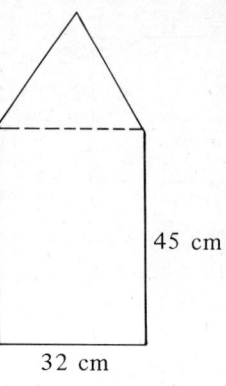

EXERCISES 2-1-E

Determine the perimeter of each of the given figures.

1. (All intersecting lines are ⊥.)

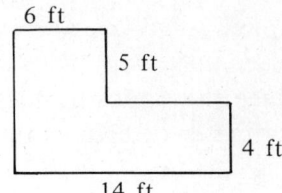

2. (The combination of a square and isosceles triangle.)

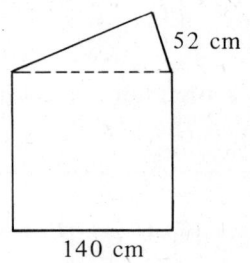

2-1-13 The solution of many important applications can be found by determining the perimeter of an appropriate geometric figure. Applications of this type are shown in the following article.

2-1-14 **EXAMPLE 2-1-F**

What is the cost of weather-stripping three rectangular windows, each 2.5 ft wide and 3 ft high, at 12 cents per foot?

By multiplying the perimeter of a window by 12 cents, we find the cost of weather-stripping one window. Then the total cost is found by multiplying by 3, since there are three windows. The perimeter of one window is

$p = 2(2.5 \text{ ft}) + 2(3 \text{ ft}) = 5 \text{ ft} + 6 \text{ ft} = 11 \text{ ft}.$

Multiplying by $0.12, we have

$C = 11(\$0.12) = \$1.32$

as the cost for weather-stripping one window. Thus, the total cost is 3($1.32) = $3.96.

EXERCISES 2-1-F

1. How much fencing is required to enclose a rectangular field 120 m long and 105 m wide?

2. The frame of the top of a card table is made from metal tubing which costs 7 cents per foot. What is the cost of the frames for two tabletops which are 33 in. square?

2-1-15 In this section we have defined the perimeter of a geometric figure as the distance around the figure. Also, we developed some formulas based upon the definition. The following exercises provide an opportunity to review how perimeters are determined.

2-1-16 **EXERCISES 2-1-Section**

In Exercises 1-20, find the perimeters of the indicated geometric figures.

1.

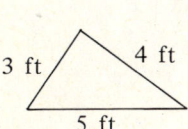

2.

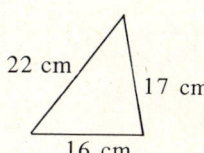

3.

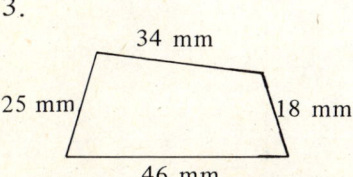

4.

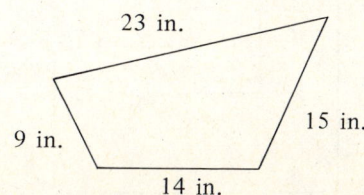

5. triangle: $a = 320$ m, $b = 278$ m, $c = 298$ m
6. triangle: $a = 52$ yd, $b = 49$ yd, $c = 64$ yd
7. quadrilateral: $a = 16.5$ in., $b = 17.3$ in., $c = 21.8$ in., $d = 29.2$ in.
8. quadrilateral: $a = 6.92$ cm, $b = 8.26$ cm, $c = 9.93$ cm
9. equilateral triangle: $s = 64.6$ mm
10. equilateral triangle: $s = 128$ ft
11. isosceles triangle: $s = 15.3$ in., $b = 26.5$ in.
12. isosceles triangle: $s = 36.2$ cm, $b = 12.5$ cm
13. square: $s = 0.65$ m
14. square: $s = 2.36$ yd
15. rhombus: $s = 0.15$ mi
16. rhombus: $s = 178$ mm
17. parallelogram: $a = 47.2$ cm, $b = 36.8$ cm
18. parallelogram: $a = 1.69$ in., $b = 1.46$ in.
19. rectangle: length = 68.7 ft, width = 46.6 ft
20. rectangle: length = 4.57 m, width = 0.97 m

In Exercises 21-28, find the perimeters of the indicated geometric figures. Unless otherwise noted, intersecting lines are perpendicular. (Dashed lines are used to identify figures, but are not part of the figure of which the perimeter is to be found.)

21.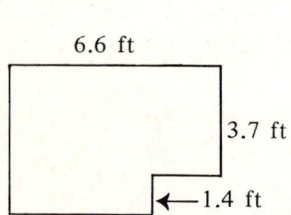

22.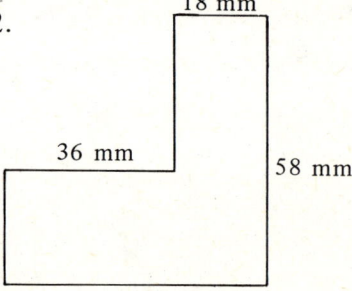

23.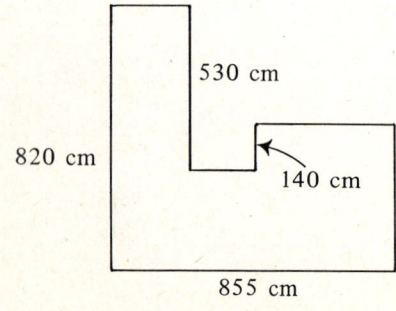

24.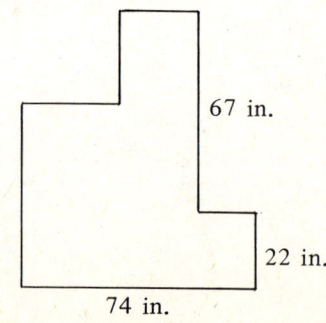

25.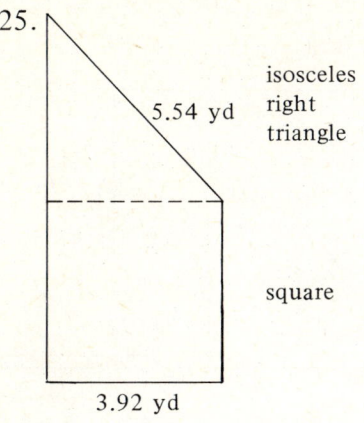

26.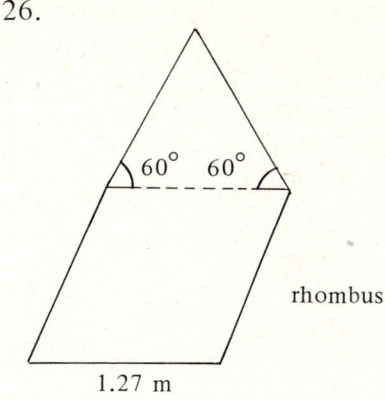

27.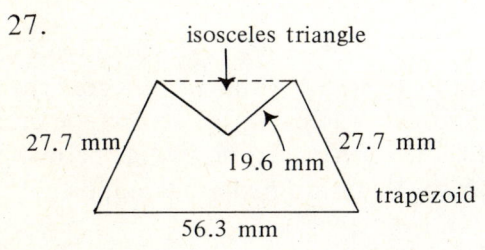

28.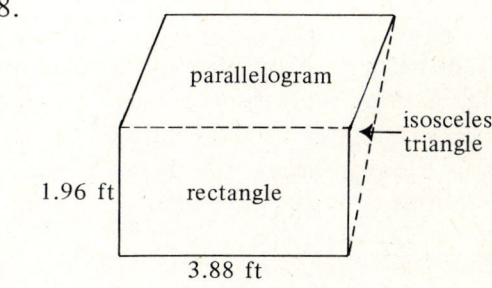

In Exercises 29-32, solve the given problems.

29. Rug binding costs 24 cents per foot. What is the cost of the binding for a rectangular rug 24 ft by 15 ft?

30. The floor of a room is in the shape of a square 9.2 m on a side. The room has two doors 0.9 m wide. How many meters of floor molding are required for this room?

31. A machine part is in the shape shown in the figure at the right. A metal strip is to be attached along the perimeter of this part. How long should the strip be?

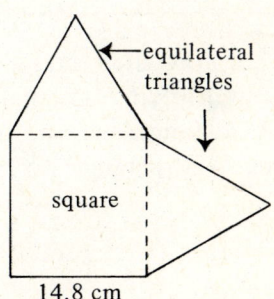

32 Basic Geometric Measures and Properties

32. Heating pipe is to be placed within the outside wall of the room shown in the figure at the right. If pipe costs 96 cents per foot, what is the cost of the pipe for this room?

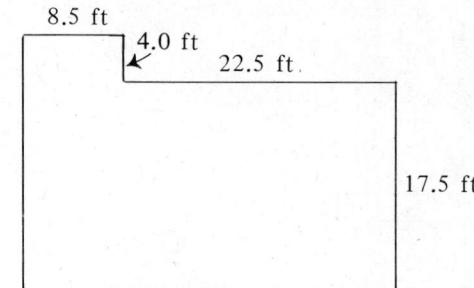

2-2 AREA

2-2-1 Another basic measure of a plane geometric figure is its *area*. Although the concept of area is primarily intuitive, it is easily defined and calculated for the basic geometric figures. Area gives us a measure of the surface of a geometric figure, just as perimeter gives us the measure of the distance around it. Also, there are numerous important applications of area.

2-2-2 In finding the area of a geometric figure we are finding the number of squares, one unit on a side, which are required to cover the surface of the figure. In the rectangle at the right, we see that 12 squares, each 1 cm on a side, are required to cover the surface. Thus, we say that the area of the rectangle is 12 sq cm.

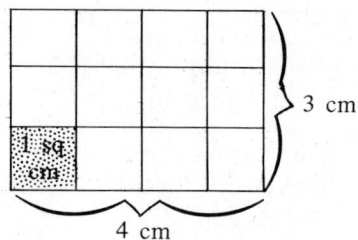

2-2-3 If we note the rectangle in Article 2-2-2, we see that its area may be determined by multiplying its length by its width. Since the area of any rectangle may be found in this way, we define the area of a rectangle to be

$$A = lw,$$

where *l* and *w* must be measured in the same unit of length, and *A* is in square units of length. (We could also have used *a* and *b* for the dimensions, which would lead to $A = ab$.)

2-2-4   Since a square is a rectangle with all sides equal in length, its area is

$$A = s^2,$$

where s is the length of one of the sides.

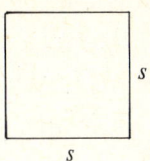

2-2-5

EXAMPLE 2-2-A

(a) The area of a rectangle of length 8 ft and width 6 ft is

$$A = (8 \text{ ft})(6 \text{ ft}) = 48 \text{ sq ft.}$$

(b) The area of a square of side 16 mm is

$$A = (16 \text{ mm})^2 = 256 \text{ sq. mm.}$$

EXERCISES 2-2-A

Find the areas of the indicated figures.

1. rectangle: length = 16 in., width = 10 in.
2. rectangle: l = 20 m, w = 14 m
3. square: side = 120 cm
4. square: s = 6.5 ft

2-2-6   We may use the definition of the area of a rectangle to determine the method for finding the area of a parallelogram. Considering the parallelogram on the left, if we draw a perpendicular line as shown (the dashed line—labeled h), a triangular area is formed.

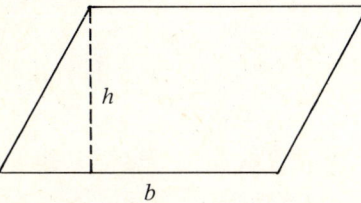 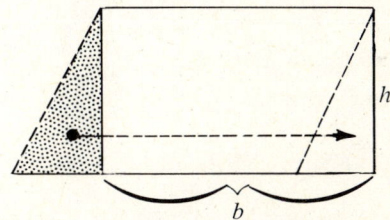

It may be thought of as being moved to the right side of the parallelogram, as shown in the right figure. A rectangle of length b and width h is thereby formed. The area of the original parallelogram and rectangle are equal. Thus, the area of the parallelogram is

$$A = bh,$$

where h is called the *altitude* (or *height*) of the parallelogram. It is the length of the perpendicular line drawn from one side to the opposite *base*, b.

34 Basic Geometric Measures and Properties

2-2-7   **EXAMPLE 2-2-B**

(a) The area of a parallelogram for which $h = 9$ m and $b = 11$ m is

$A = (11 \text{ m})(9 \text{ m}) = 99$ sq m.

(b) To find the area of the parallelogram shown below, we need only the height, 6.5 in., and base, 8.0 in. The length of the other side, 7.0 in., is not used. (It would be used to find the perimeter.) Thus,

$A = (8.0 \text{ in.})(6.5 \text{ in.})$
$ = 52$ sq in.

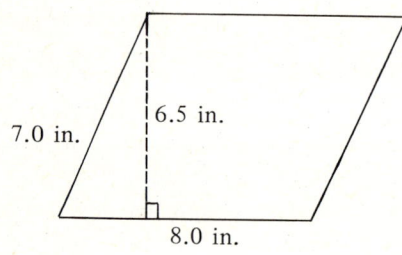

EXERCISES 2-2-B

Find the areas of the indicated parallelograms.

1. $h = 3.5$ ft, $b = 4.0$ ft
2. $h = 28$ mm, $b = 20$ mm
3.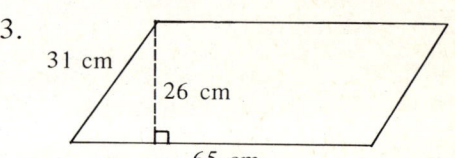
4.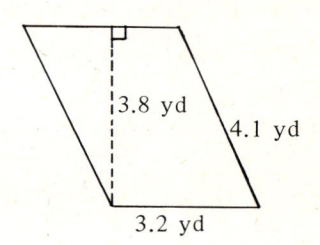

2-2-8   If we draw a diagonal in a parallelogram, we divide it into two equal triangles, as shown in the figure.

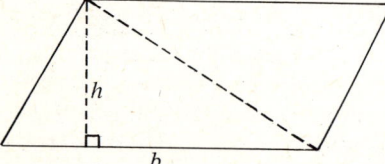 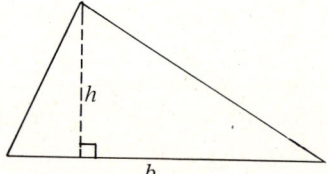

Since the area of the triangle is one-half the area of the parallelogram, we have

$$A = \frac{1}{2}bh$$

as the area of the triangle. Here, h is the *altitude* (or *height*) of the triangle. It is the length of the perpendicular line from a vertex to the base of the triangle.

2-2-9 **EXAMPLE 2-2-C**

(a) The area of a triangle for which $h = 16$ in. and $b = 12$ in. is

$$A = \frac{1}{2}(16 \text{ in.})(12 \text{ in.}) = 96 \text{ sq in.}$$

(It might be noted that $\frac{1}{2}bh$ and $\frac{bh}{2}$ are the same.)

(b) The area of the right triangle shown is

$$A = \frac{1}{2}(1.8 \text{ cm})(1.7 \text{ cm})$$
$$= 15.3 \text{ sq cm.}$$

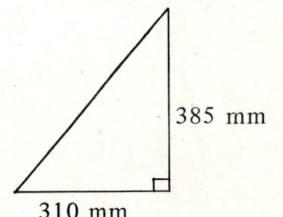

Since the legs of a right triangle are perpendicular, either can be considered as the base and the other the altitude.

EXERCISES 2-2-C

Find the areas of the indicated triangles.

1. $h = 3.9$ m, $b = 4.2$ m
2. $h = 29$ ft, $b = 18$ ft
3.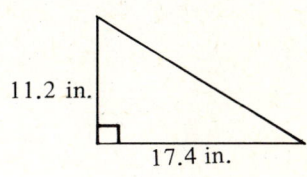
4.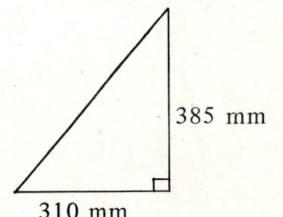

(triangle with 11.2 in. and 17.4 in.)

(triangle with 385 mm and 310 mm)

2-2-10 When we join opposite vertices of a trapezoid with a diagonal, two triangles are formed, as is seen in the figure below. The area of the lower triangle is $\frac{1}{2}b_1 h$ and the area of the upper triangle is $\frac{1}{2}b_2 h$. The sum of these areas is

$$\frac{1}{2}b_1 h + \frac{1}{2}b_2 h = \frac{1}{2}h(b_1 + b_2).$$

Since the sum is the area of the trapezoid, we have

$$A = \frac{1}{2}h(b_1 + b_2)$$

for the area of the trapezoid.

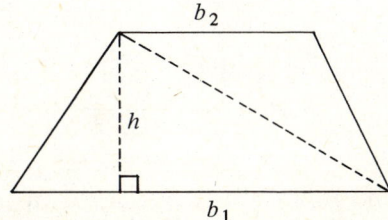

2-2-11 EXAMPLE 2-2-D

(a) The area of the trapezoid for which $h = 32$ mm, $b_1 = 14$ mm, and $b_2 = 41$ mm is

$A = \frac{1}{2}(32 \text{ mm})(14 \text{ mm} + 41 \text{ mm})$

$ = \frac{1}{2}(32 \text{ mm})(55 \text{ mm}) = 880$ sq mm.

(b) The area of the trapezoid shown is

$A = \frac{1}{2}(12 \text{ ft})(8 \text{ ft} + 22 \text{ ft})$

$ = \frac{1}{2}(12 \text{ ft})(30 \text{ ft})$

$ = 180$ sq ft.

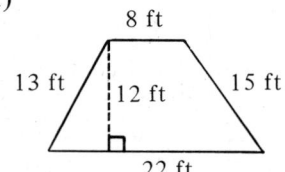

Note that the lengths of the nonparallel sides were not used in the calculation.

EXERCISES 2-2-D

Find the areas of the indicated trapezoids.

1. $h = 6$ in., $b_1 = 15$ in., $b_2 = 9$ in.
2. $h = 1.6$ cm, $b_1 = 3.4$ cm, $b_2 = 2.7$ cm
3.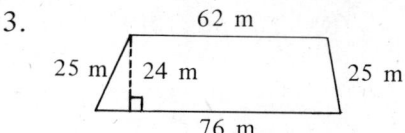
4.

3.4 ft

1.8 ft — 1.2 ft

4.8 ft

2-2-12

The determination of area is important in many types of applications involving geometric figures. Consider the illustrations in the following article.

2-2-13 EXAMPLE 2-2-E

How much will it cost to carpet the area shown with carpeting costing $12.50/sq m? (The area is a combination of a rectangle and right triangle.)

To determine the cost, we must find the total area and multiply by $12.50. The total area is the area of the rectangle plus the area of the triangle. The rectangle is 1.9 m by 4.6 m. The base of the triangle is 4.6 m − 2.5 m = 2.1 m, and its height is 0.6 m. Thus,

$A = (1.9 \text{ m})(4.6 \text{ m}) + \frac{1}{2}(0.6 \text{ m})(2.1 \text{ m}) = 9.37$ sq m.

The cost is

$C = (9.37 \text{ sq m})(\$12.50/\text{sq m}) = \$117.13.$

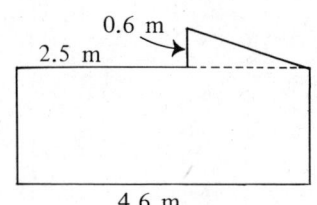

EXERCISES 2-2-E

1. What is the area of a rectangular wall 14.0 ft by 8.0 ft in which there is a square window 3.5 ft on a side?

2. A special coating which costs $0.85/sq cm is to be put on the surface shown. What is the cost of the coating?

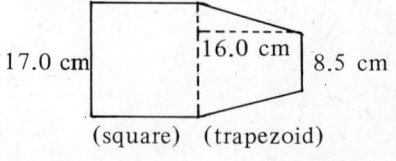

(square) (trapezoid)

2-2-14 In this section we have defined the area of a geometric figure. Also we developed specific formulas for the areas of a rectangle, square, parallelogram, triangle, and trapezoid. The following exercises provide an opportunity to review the calculation of areas.

2-2-15 **EXERCISES 2-2-Section**

In Exercises 1-16, find the areas of the indicated geometric figures.

1. rectangle: $l = 60$ cm, $w = 45$ cm
2. rectangle: $l = 152$ ft, $w = 85$ ft
3. square: $s = 7.6$ in.
4. square: $s = 0.16$ km
5. parallelogram: $b = 72$ mm, $h = 34$ mm
6. parallelogram: $b = 1.5$ yd, $h = 1.2$ yd
7. rhombus: side = 16.5 in., altitude = 6.4 in.
8. rhombus: side = 240 cm, altitude = 150 cm
9. triangle: $b = 0.75$ m, $h = 0.64$ m
10. triangle: $b = 64.0$ in., $h = 14.5$ in.
11. right triangle: legs: 16.5 ft and 28.8 ft
12. right triangle: legs: 396 mm and 250 mm
13. trapezoid: $h = 0.012$ km, $b_1 = 0.025$ km, $b_2 = 0.018$ km
14. trapezoid: $h = 1.23$ yd, $b_1 = 3.74$ yd, $b_2 = 2.36$ yd
15. trapezoid shown
16. trapezoid shown

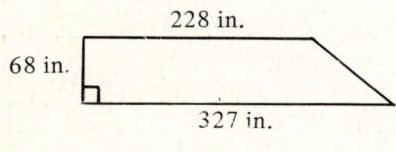
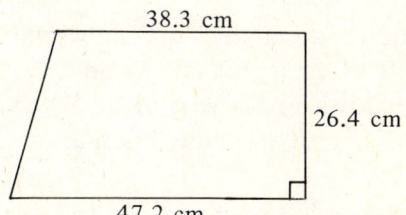

In Exercises 17-20, determine (a) the area and (b) the perimeter of the geometric figures shown.

17. (parallelogram)   18. (parallelogram)   19.   20.

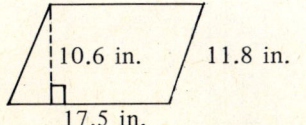
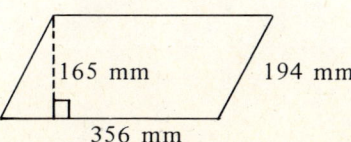
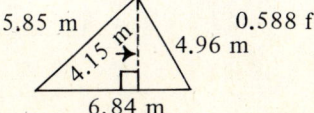
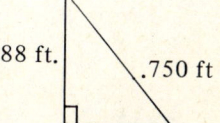

38 Basic Geometric Measures and Properties

In Exercises 21-24, find the areas of the indicated geometric figures. (Dashed lines are used to identify figures and altitudes, but are not part of the figure.)

21. (parallelogram and rectangle)

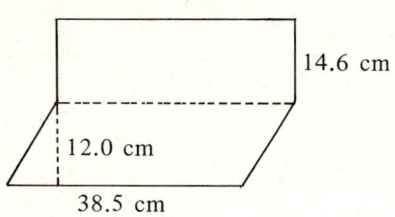

22. (square and trapezoid)

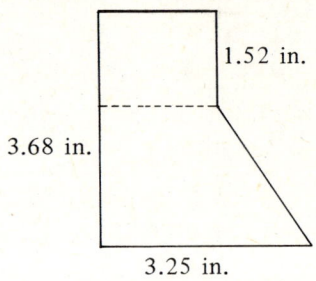

23. (triangle and trapezoid)

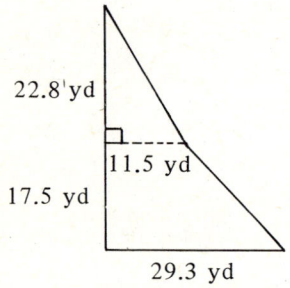

24. (square and triangles)

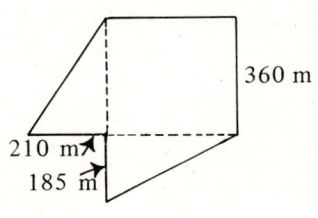

In Exercises 25-28, solve the given problems.

25. A cabinet has three rectangular shelves, each 14.5 in. by 24.0 in. What is the area of shelf space available in the cabinet?

26. A rectangular walk, 2.5 ft wide, is to be placed around a rectangular garden 36.5 ft by 67.5 ft. What is the cost of a cement walk if the cost is $1.60/sq ft?

27. The figure at the right shows the side of a house. It is a rectangle and triangle, with rectangular windows. At a cost of $2.50/sq m, how much will it cost to paint this side of the house?

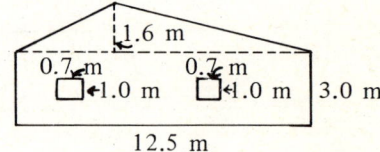

28. A fence is to be made of trapezoidal slats as shown in the figure at the right. What is the area of the fence if it is placed along a distance of 13.34 m? (100 cm = 1 m)

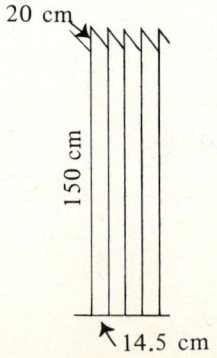

2-3 THE PYTHAGOREAN THEOREM

2-3-1 In a right triangle, if two of the sides are known, the third side may be determined by use of the *Pythagorean theorem*. In this section we develop this very important theorem and discuss some of its many applications in technology.

2-3-2 The Pythagorean theorem states that

> in a right triangle, the square of the length of the hypotenuse equals the sum of the squares of the lengths of the other two sides.

Referring to the triangle shown below, the equation for the Pythagorean theorem is

$$a^2 + b^2 = c^2.$$

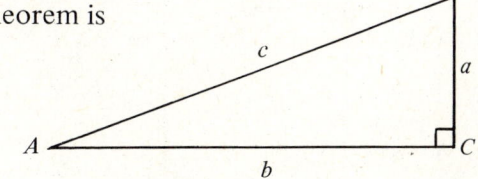

2-3-3 We can prove the Pythagorean theorem by noting the areas of the squares and triangles in the figure. A square of side c is inscribed in a square of side $a + b$. The area of the outer square $(a + b)^2$ minus the area of the four triangles of sides a, b, and c (the area of each of these triangles is $\frac{1}{2} ab$) equals the area of the inner square. This leads to

$$(a + b)^2 - 4(\tfrac{1}{2} ab) = c^2$$
$$a^2 + 2ab + b^2 - 2ab = c^2$$
$$a^2 + b^2 = c^2.$$

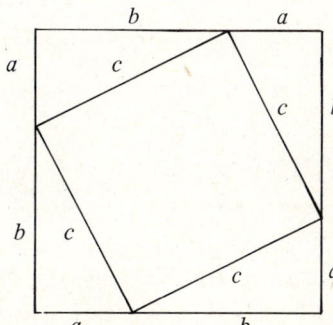

40   Basic Geometric Measures and Properties

2-3-4   In working with the Pythagorean theorem, it is generally necessary to determine square roots. It is also necessary to consider *rounding off* and *significant digits* when evaluating these square roots. If these terms are not familiar, a brief discussion of rounding off and significant digits is found in Appendix A, and a table of squares and square roots along with illustrations of its use is found in Appendix B. Other means of determining square roots are logarithms, the slide rule, or certain electronic calculators. *Please note that after this point all rounded off calculations have been rounded off to three significant digits, except where noted.*

2-3-5   **EXAMPLE 2-3-A**

(a) In the triangle shown at the right, we are to find the hypotenuse c. From the Pythagorean theorem, we have

$$6^2 + 8^2 = c^2$$

or

$$c^2 = 36 + 64 = 100.$$

Since $c^2 = 100$, $c = \sqrt{100}$. Thus, $c = 10$.

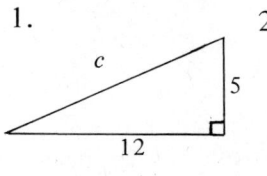

(b) In finding the hypotenuse c in the triangle at the right, we have

$$9^2 + 4^2 = c^2$$

or

$$c^2 = 81 + 16 = 97.$$

Thus,

$$c^2 = \sqrt{97} = 9.85.$$

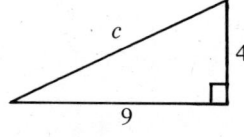

EXERCISES 2-3-A

Find the hypotenuse (to three significant digits, if rounding off is necessary) in each of the given right triangles by use of the Pythagorean theorem.

1.   2.

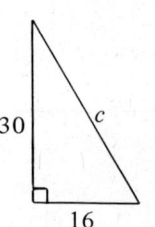

3.   4.

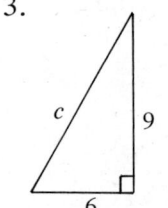 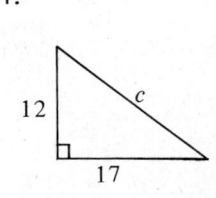

2-3-6   In each of the illustrations of the previous example, we were required to find the hypotenuse of a right triangle. We can also determine one of the legs of a right triangle if the other leg and the hypotenuse are known. The following example illustrates this use of the Pythagorean theorem.

2-3-7 EXAMPLE 2-3-B

(a) In the right triangle shown at the right, we are to find leg b. From the Pythagorean theorem, we have

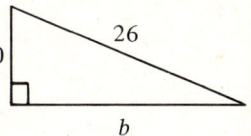

$$10^2 + b^2 = 26^2$$

or

$$b^2 = 26^2 - 10^2 = 676 - 100 = 576.$$

Thus,

$$b = \sqrt{576} = 24.$$

(b) In a right triangle with legs a and b and hypotenuse c, $b = 3.7$ and $c = 5.4$. To find a, we have

$$a^2 + 3.7^2 = 5.4^2$$

or

$$a^2 = 5.4^2 - 3.7^2 = 29.16 - 13.69 = 15.47.$$

Thus,

$$a = \sqrt{15.47} = 3.93.$$

EXERCISES 2-3-B

Find the unknown side (to three significant digits, if rounding off is necessary) in each of the given right triangles by use of the Pythagorean theorem. In problems 3 and 4, a and b are the legs and c is the hypotenuse.

1.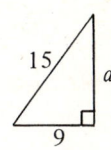

2.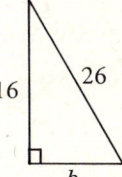

3. $a = 1.8$, $c = 2.4$, find b

4. $b = 230$, $c = 350$, find a

2-3-8 There are numerous important applications of the Pythagorean theorem in nearly all branches of mathematics, science, and technology. The following two examples illustrate certain of these applications, and others are found in the exercises at the end of the section.

2-3-9 EXAMPLE 2-3-C

How long must a guy wire be if it is attached 25 m up on an antenna and on the ground 30 m from the foot of the antenna?

First, we interpret the statement of the problem by drawing a right triangle with the vertical side representing the antenna and the horizontal side representing the ground. The hypotenuse represents the guy wire. The length x of the guy wire is found by use of the Pythagorean theorem.

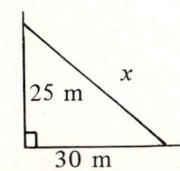

$$25^2 + 30^2 = x^2; \quad x^2 = 625 + 900 = 1525$$

$$x = \sqrt{1525} = 39.1$$

Thus, the guy wire is 39.1 m long.

EXERCISES 2-3-C

1. A rectangular television screen is 15.7 in. wide and 11.7 in. high. How long is the diagonal distance from one corner to the opposite corner?

2. A ship travels 17.5 km south from a port and then turns and travels an additional 28.0 km due east. How far from the port is the ship?

2-3-10 Up to the example in the previous article we had included the units of measurement in the equation and algebraic work. Technically this should always be done, but it can be cumbersome, and recognizing that all lengths are in meters, we know the result is in meters. Therefore, from this point on, where the units of measurement of the given parts and result are specifically known, we will not include them in the equation or algebraic work.

2-3-11 **EXAMPLE 2-3-D**

The base of a 20-foot ladder is 8.0 ft from a wall. How far up on the wall does the ladder reach?

From the statement of the problem we set up the diagram at the right. We are to find the vertical distance x.

$$8.0^2 + x^2 = 20^2$$
$$x^2 = 400 - 64 = 336$$
$$x = \sqrt{336} = 18.3$$

Thus, the ladder reaches 18.3 ft up the wall.

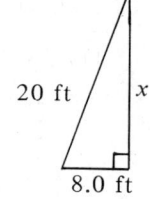

EXERCISES 2-3-D

1. A jet travels 7.5 km while gaining altitude at a constant rate. If it traveled between horizontal points 5.8 km apart, what was its gain in altitude?

2. How far from a 75-ft tower must an observer stand such that his feet are 125 ft from the top of the tower?

2-3-12 In this section we have developed the very important Pythagorean theorem. The basic statement of the theorem is found in Article 2-3-2. We also determined how it is used to find one side of a right triangle when the other two are known. Applications of the use of the Pythagorean theorem were also discussed. The following exercises provide an opportunity to review the use of the Pythagorean theorem.

2-3-13 **EXERCISES 2-3-Section**

In Exercises 1-12, find the indicated sides of the right triangle shown at the right. Where necessary, round off results to three significant digits.

1. $a = 3$, $b = 4$, $c = ?$
2. $a = 7$, $b = 24$, $c = ?$
3. $a = 20$, $c = 29$, $b = ?$
4. $a = 8$, $c = 17$, $b = ?$
5. $b = 36$, $c = 39$, $a = ?$
6. $b = 15$, $c = 25$, $a = ?$
7. $a = 7.50$, $b = 8.30$, $c = ?$
8. $a = 107$, $b = 215$, $c = ?$
9. $a = 73.7$, $c = 86.1$, $b = ?$
10. $a = 0.962$, $c = 0.995$, $b = ?$
11. $b = 90.0$, $c = 106$, $a = ?$
12. $b = 16.5$, $c = 42.4$, $a = ?$

In Exercises 13-20, solve the given problems by the appropriate use of the Pythagorean theorem.

13. What is the diagonal distance between corners of a rectangular room 12.5 ft wide and 17.0 ft long?

14. A 22.0-meter-high tree casts a shadow 15.6 m long. How far is it from the top of the tree to the tip of the shadow?

15. An observer is 550 m from the launch pad of a rocket. After the rocket has ascended vertically to a point which is 750 m from the observer, how far has it ascended?

16. A 35.0-ft-high pole is to be supported by a 58.0-ft guy wire attached at its top. How far from the base of the pole is the guy wire attached on the ground?

17. A searchlight is 520 ft from a wall, and its beam reaches a point 38.0 ft up on the wall. What is the length of the beam?

18. A man rows across a river that is 180 m wide. The current carries him downstream 20 m from the point directly across from his starting point. How far did he actually travel?

19. A stairway railing is 5.5 m long and extends along a stairway between floors which are 2.8 m apart. What is the horizontal distance between the ends of the railing?

20. A motorist travels 25.0 mi due east of a town. How far due north must he now travel to be 28.0 mi from the town?

2-4 SIMILAR TRIANGLES

2-4-1   In our discussion of geometry to this point we have studied several important basic measures of geometric figures, such as perimeter and area, which deal with the actual size of the figure. In this section we shall consider the properties of triangles that are of the same basic shape although not necessarily of the same size. Such triangles are called *similar* triangles.

2-4-2   There are two important basic properties of triangles which are similar. These properties are:

1. *corresponding angles* are equal, and
2. *corresponding sides* are proportional.

Corresponding sides and corresponding angles—one each for two triangles—are those which have the same relative position within each triangle.

2-4-3   **EXAMPLE 2-4-A**

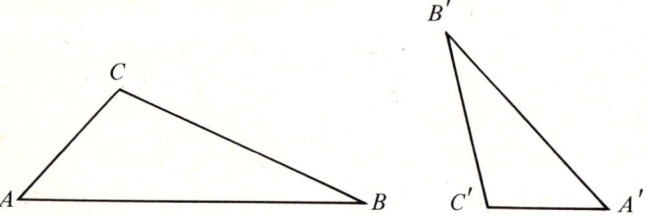

The triangles shown in the figure above have been lettered so that corresponding angles have the same letter. That is, ∠A and ∠A' are corresponding angles, as are ∠B and ∠B', and ∠C and ∠C'. (The symbol ' is read as "prime.") The corresponding sides are AB and A'B', BC and B'C', and AC and A'C'. Even though the triangles are not drawn so that corresponding parts are in the same position relative to the page, the corresponding parts are in the same relative position within each triangle. The largest angle is between the shortest and next shortest sides, and so on.

EXERCISES 2-4-A

For each pair of triangles, identify the corresponding angles and the corresponding sides.

1.
2.
3.
4.

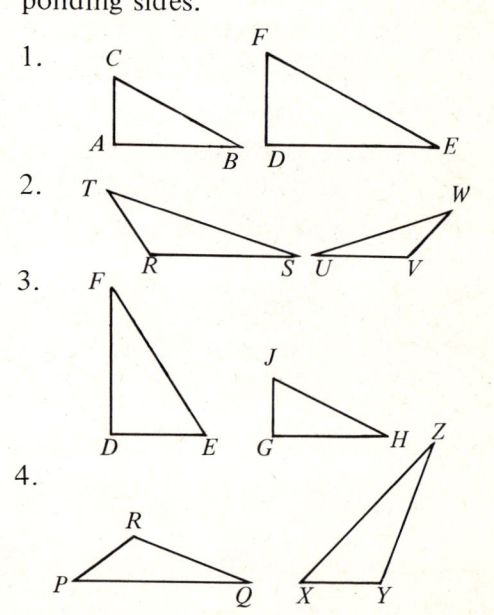

2-4-4   **EXAMPLE 2-4-B**

(a) The triangles in Example 2-4-A are similar. Therefore, the corresponding angles are equal; this means that $\angle A = \angle A'$, $\angle B = \angle B'$, and $\angle C = \angle C'$.

(b) The fact that the corresponding sides are proportional means that the ratio of any side to its corresponding side is always the same. That is,
$$\frac{AB}{A'B'} = \frac{BC}{B'C'} = \frac{AC}{A'C'}.$$

(c) In using the equation in part (b) we must have one pair of corresponding sides known along with one other side. For example, given that $AB = 12$, $A'B' = 8$, and $BC = 9$, we have
$$\frac{12}{8} = \frac{9}{B'C'}, \quad 12\,B'C' = 72, \quad B'C' = 6.$$
From this information we do not use the fraction involving AC and $A'C'$ nor can we find either AC or $A'C'$.

EXERCISES 2-4-B

The triangles in each exercise of Exercises 2-4-A are similar. For the three given sides in each exercise below, find the indicated side.

1. $AB = 4$, $AC = 2$, $DE = 8$, find DF.
2. $RS = 9$, $ST = 12$, $UV = 6$, find UW.
3. $FE = 24$, $GH = 15$, $JH = 18$, find FD.
4. $PR = 4$, $XY = 7$, $YZ = 14$, find RQ.

2-4-5   If the corresponding angles and the corresponding sides of two triangles are *equal*, then the two triangles are said to be *congruent*. For example, a right triangle with legs of 2 in. and 4 in. is congruent to any other right triangle with legs of 2 in. and 4 in. However, it is similar to any right triangle with legs 5 in. and 10 in. since the ratios of corresponding sides are equal.

2-4-6   We have seen that similar triangles have two basic properties. These are (1) corresponding angles are equal, and (2) corresponding sides are proportional. The use of similar triangles in technical applications generally is in finding a given side, knowing a pair of corresponding sides and another side. Thus, we make use of the property that corresponding sides are proportional. The property that corresponding angles are equal is often used in determining that two triangles are similar.

2-4-7 EXAMPLE 2-4-C

In the figures below, $\angle A = \angle D$, $\angle B = \angle E$, and $\angle C = \angle F$. Thus, we know that $\triangle ABC \sim \triangle DEF$ (the symbol $\sim$ means "similar to"). This, in turn, tells us that corresponding sides are proportional, or

$$\frac{AC}{DF} = \frac{CB}{FE} = \frac{BA}{ED}.$$

If we now are given $AC = 6$ in., $DF = 4$ in., $FE = 3$ in., and $ED = 2$ in., we have

$$\frac{6}{4} = \frac{CB}{3} = \frac{BA}{2}.$$

Therefore,

$$CB = \frac{6(3)}{4} = \frac{9}{2} \text{ in.,}$$

$$BA = \frac{6(2)}{4} = 3 \text{ in.}$$

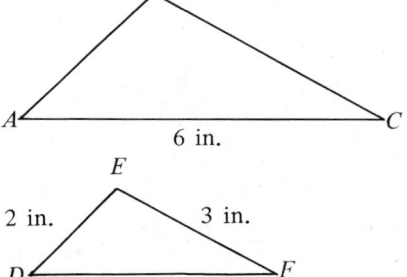

EXERCISES 2-4-C

In the figures below, $\angle P = \angle S$, $\angle Q = \angle T$ and $\angle R = \angle U$.

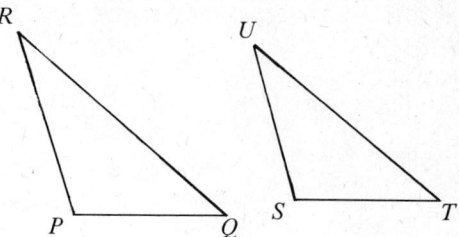

Given $ST = 8$, $TU = 16$, $SU = 12$, and $PQ = 14$, determine:

1. QR
2. PR

2-4-8 EXAMPLE 2-4-D

In constructing a metal structure support in the form of $\triangle ABC$ as shown below, it is deemed necessary to strengthen the support with an added brace DE which is parallel to BC. How long must the brace DE be if $AB = 20$ in., $AD = 14$ in., and $BC = 25$ in.?

Since $DE \parallel BC$, from corresponding angles we see that $\angle CBA = \angle EDA$ and $\angle BCA = \angle DEA$. Since $\angle A$ is common to both triangles, we have $\triangle ABC \sim \triangle ADE$. Thus,

$$\frac{AD}{AB} = \frac{DE}{BC} \text{ or } \frac{14}{20} = \frac{DE}{25}.$$

This leads to

20 DE = 350, or DE = 17.5.

Therefore, the brace should be 17.5 in. long

EXERCISES 2-4-D

1. In the structure shown in Example 2-4-D, find length AE if $AC = 30$ in.

2. A one-meter stick is placed vertically in the shadow of a vertical pole such that the ends of their shadows are at the same point. If the shadow of the meter stick is 80 cm long and that of the pole is 280 cm long, how high is the pole?

2-4-9   One of the most practical uses of similar geometric figures (including similar triangles) is that of *scale drawings*. Maps, charts, blueprints, and most drawings which appear in books are familiar examples of scale drawings. Actually there have been numerous scale drawings used in this module in the previous articles.

2-4-10 In any scale drawing, all distances are drawn a certain ratio of the distances they represent. If the scale of the drawing is known, it is not necessary to actually use similar triangles, although the procedure is similar in that a proportion is set up. [If the scale is not known (which is unusual), the similar triangles which can be used are those of the drawing and the actual distances, although we can only represent the actual distances.]

2-4-11 **EXAMPLE 2-4-E**

(a) In drawing a map of the area shown in the figure below, a scale of 1 cm = 200 km is used. In measuring the distance between Chicago and Toronto on the map, we find it to be 3.5 cm.

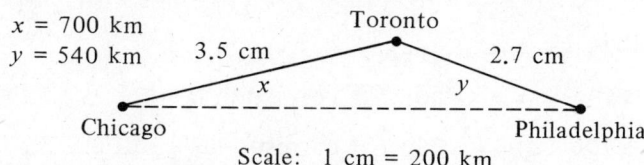

Therefore, the actual distance x between Chicago and Toronto is found from the proportion
$$\frac{x}{3.5 \text{ cm}} = \frac{200 \text{ km}}{1 \text{ cm}}, \text{ or } x = 700 \text{ km}.$$

(b) If we did not have the scale, but did know that the distance between Chicago and Toronto is 700 km, then by measuring distances on the map between Chicago and Toronto (3.5 cm) and between Toronto and Philadelphia (2.7 cm) we could find the distance between Toronto and Philadelphia from the following proportion, found by use of similar triangles:
$$\frac{700 \text{ km}}{3.5 \text{ cm}} = \frac{y}{2.7 \text{ cm}}$$
$$y = \frac{2.7(700)}{3.5} = 540 \text{ km}$$

EXERCISES 2-4-E

In drawing the structure shown in the figure below, a scale of 1 in. = 4 ft is used.

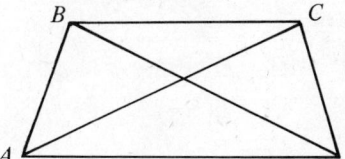

1. By use of the scale and measuring BC, find the distance between B and C on the actual structure.

2. By knowing that the distance between A and B is 3 ft, and measuring BC, find the distance between B and C on the actual structure. (Do not use the scale.)

2-4-12 In this section we discussed similar triangles and their properties. The two basic properties of similar triangles are (1) corresponding angles are equal, and (2) corresponding sides are proportional. We have also discussed some applications, including the related application of scale drawings. The following exercises provide an opportunity to review similar triangles and applications.

2-4-13 EXERCISES 2-4-Section

In Exercises 1-4, for the given pairs of triangles, identify the corresponding angles and corresponding sides to the given angles and sides. The triangles in each of the figures are similar.

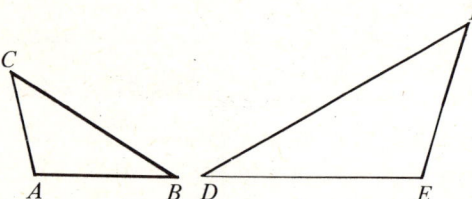
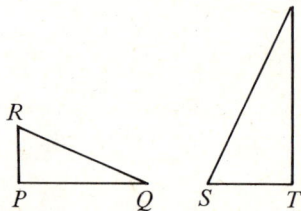

           Figure 1   Figure 2

1. In Figure 1, $\angle A$ corresponds to _____ , side AC corresponds to _____ .
2. In Figure 1, $\angle F$ corresponds to _____ , side DF corresponds to _____ .
3. In Figure 2, $\angle Q$ corresponds to _____ , side RP corresponds to _____ .
4. In Figure 2, $\angle S$ corresponds to _____ , side UT corresponds to _____ .

In Exercises 5-8, for the given sides of the triangles of Figures 1 and 2, find indicated sides.

    In Figure 1: $AB = 5$, $BC = 7$, $AC = 4$, and $FE = 8$.
    In Figure 2: $RP = 6$, $PQ = 8$, $RQ = 10$, and $ST = 9$.

5. In Figure 1, find DE.   6. In Figure 1, find FD.
7. In Figure 2, find TU.   8. In Figure 2, find SU.

In Exercises 9-12, use $\triangle ACE$ at the right. In this figure $BD \parallel AE$, $AE = 18$, and $DB = 6$.

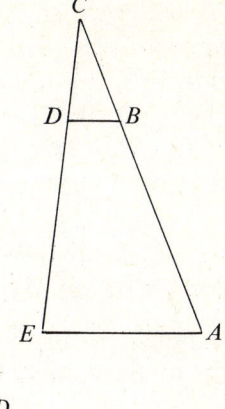

9. If $DC = 7$, find CE.

10. If $CA = 27$, find CB.

11. If $CB = 10$, find BA.

12. If $CE = 24$, find DE.

In Exercises 13-16, use $\triangle ABC$ at the right. In the figure $AC \perp BC$ and $CD \perp AB$.

13. Given that $\triangle ABC \sim \triangle ACD$, find AB if $AD = 9$ and $AC = 12$.

14. Given that $\triangle ABC \sim \triangle CBD$, find BD if $BC = 6$ and $AB = 9$.

15. Given that $\triangle ACD \sim \triangle CBD$, find BD if $CD = 9$ and $AD = 27$.

16. Show that $\triangle ABC \sim \triangle ACD$. (Show that corresponding angles are equal. Each triangle has a right angle, and they have a common angle.)

In Exercises 17-24, solve the given problems by the appropriate use of similar triangles and proportions.

17. On level ground, a tree casts a shadow 36 ft long. At the same time, a pole 9 ft high casts a shadow 12 ft long. How high is the tree?

18. A 200-cm-tall man casts a shadow 160 cm long. At the same time his son casts a shadow 120 cm long. How tall is the son?

19. A good approximation of the height of a tree can be made by following the procedure suggested in the figure at the right.
By measuring DE, AE, and BC (use a ruler), the length of the tree $DE + EF$ can be found. Find the height of a tree if $AB = 50$ cm, $BC = 30$ cm, $AE = 2400$ cm, and $DE = 150$ cm.

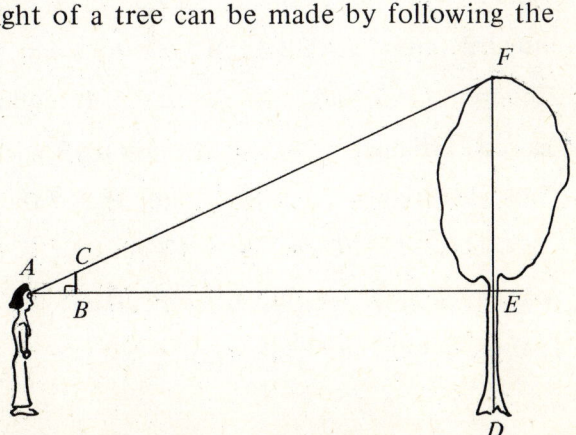

20. To find the width ED of a river, a surveyor places markers at A, B, C, and D as shown in the figure at the right. He places them such that AB ∥ ED, BC = 50 ft, DC = 300 ft, and AB = 80 ft. How wide is the river?

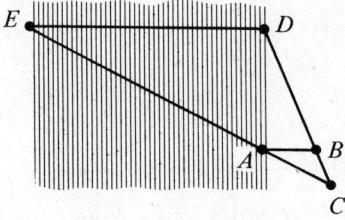

21. A rectangular painting is 6 ft high and 10 ft wide. What is the length of a diagonal of a reproduction whose width is 3 ft?

22. A rectangular object is 16 m wide. Its image, as seen through a lens, has a width of 6 cm and a length of 9 cm. What is the length of the object?

23. On the blueprint of a certain building, a hallway is 45.6 cm long. The scale is 1.2 cm = 1 m. How long is the hallway?

24. The figure at the right shows a crank-lever mechanism. From this figure determine the length AB.

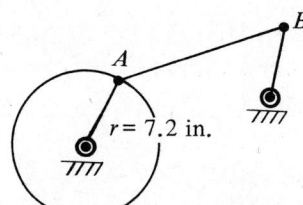

2-5 EXERCISES FOR UNIT TWO

In Exercises 1-12, find the perimeters of the indicated geometric figures.

1. triangle: $a = 17.5$ in., $b = 13.8$ in., $c = 8.9$ in.
2. quadrilateral: $a = 22.4$ cm, $b = 68.5$ cm, $c = 37.3$ cm, $d = 29.9$ cm
3. square: $s = 6.8$ m
4. equilateral triangle: $s = 14.9$ ft
5. isosceles triangle: $s = 0.38$ yd, $b = 0.53$ yd
6. isosceles triangle: $s = 8.13$ mm, $b = 7.09$ mm
7. rectangle: $l = 96$ cm, $w = 43$ cm
8. rectangle: $l = 108$ in., $w = 92$ in.
9. parallelogram: $a = 692$ ft, $b = 207$ ft
10. parallelogram: $a = 7.8$ m, $b = 6.2$ m

11.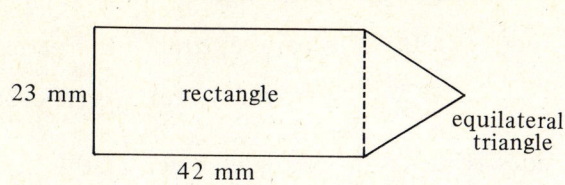

12.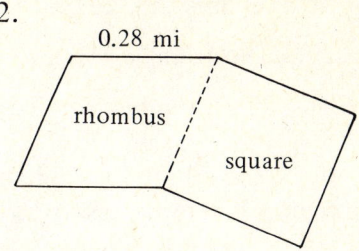

In Exercises 13-24, find the areas of the indicated geometric figures.

13. rectangle: $l = 17.5$ cm, $w = 12.8$ cm
14. square: $s = 94$ in.
15. parallelogram: $b = 2.68$ yd, $h = 1.15$ yd
16. parallelogram: $b = 172$ mm, $h = 85$ mm
17. triangle: $b = 4.68$ cm, $h = 2.05$ cm
18. triangle: $b = 68$ ft, $h = 43$ ft
19. right triangle: legs: 3.25 in. and 1.88 in.
20. right triangle: legs: 40.8 m and 19.5 m
21. trapezoid: $h = 0.016$ km, $b_1 = 0.118$ km, $b_2 = 0.067$ km
22. trapezoid: $h = 36.8$ in., $b_1 = 12.4$ in., $b_2 = 11.1$ in.
23. the figure for Exercise 11 if h of the triangle is 20 mm
24. the figure for Exercise 12 if h of the rhombus is 0.22 mi

In Exercises 25-28, find the perimeters of the indicated geometric figures.

25.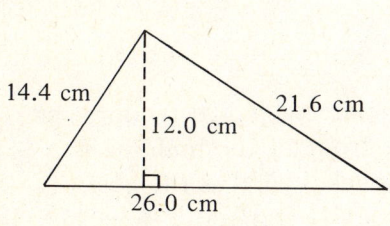

26.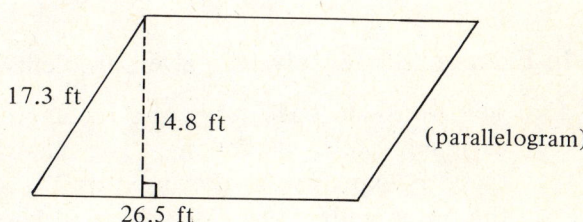

27.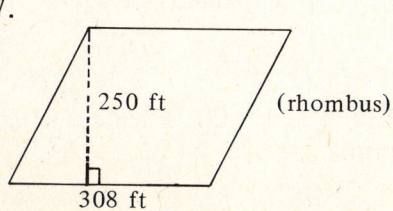

28.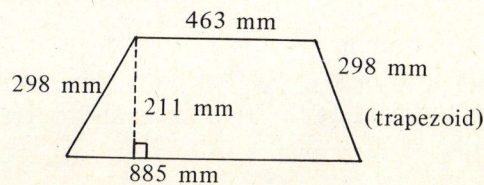

In Exercises 29-32, find the areas of the indicated geometric figures.

29. The triangle for Exercise 25
30. The parallelogram for Exercise 26
31. The rhombus for Exercise 27
32. The trapezoid for Exercise 28

In Exercises 33-40, find the indicated sides of the right triangle shown at the right. Where necessary, round off results to three significant digits.

33. $a = 9$, $b = 40$, $c = ?$
34. $a = 14$, $b = 48$, $c = ?$
35. $a = 40$, $c = 58$, $b = ?$
36. $b = 56$, $c = 65$, $a = ?$
37. $a = 6.30$, $b = 3.80$, $c = ?$
38. $a = 126$, $b = 251$, $c = ?$
39. $b = 29.3$, $c = 36.1$, $a = ?$
40. $a = 0.782$, $c = 0.885$, $b = ?$

In Exercises 41-44, for the given lengths shown in the figures, find the indicated lengths.

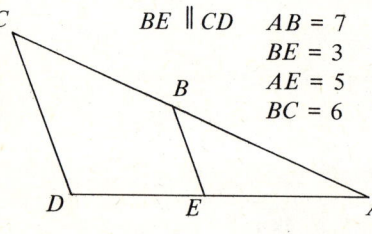
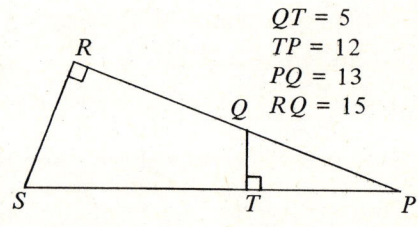

$BE \parallel CD$ $AB = 7$
$BE = 3$
$AE = 5$
$BC = 6$

$QT = 5$
$TP = 12$
$PQ = 13$
$RQ = 15$

Exercises 41 and 42

Exercises 43 and 44

41. Find CD.
42. Find DE.
43. Find RS.
44. Find ST.

In Exercises 45-56, solve the given problems.

45. The front and back walls of a rectangular building are each 110 ft long, and each side wall is 80 ft long. The material on the front of the building is twice as expensive as that used on the sides and back. What is the cost of the outer walls if the side walls and back wall cost $18 per foot?

46. What is the total cost of fencing in a right triangular plot of land, of which the legs are 45 m and 60 m, at a cost of $4 per meter?

47. A swimming pool is 1.50 m deep at one end and slopes uniformly to a depth of 3.00 m at the other end. The pool is 12.0 m long and 6.00 m wide. What is the area of the four sides and bottom of the pool?

48. A gallon of paint will cover 275 sq ft. How much paint (to the nearest tenth of a gallon) is required to paint the walls of a rectangular room 15 ft by 22 ft, given that the room has three windows 2.0 ft by 3.0 ft, two doors 3.0 ft by 6.5 ft and the walls are 8.0 ft high?

49. A roof truss is shown in the figure below. The rafters are 24.0 ft long, including a 2.0-ft overhang, and the height of the truss is 5.75 ft. Determine the length of the base of the truss.

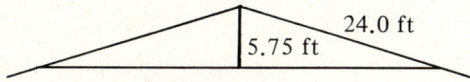

50. Determine the length of the diagonal of a square piece of land, 0.650 km on a side.

51. A drafting student makes a drawing with a scale of 2.4 cm = 8.00 m. What distance on his drawing should be used to represent 70.0 m?

52. On the blueprint of a building a rectangular hallway is 25.0 in. long and 5.25 in. wide. If the hallway is 20.0 ft long, how wide is it?

53. A 4-ft wall stands 2 ft from a building. The ends of a straight pole touch the building, the wall, and the ground 6 ft from the wall. How high up on the building does the pole touch?

54. Light is reflected from a mirror so that the angle of incidence i equals the angle of reflection r. See the figure at the right. A ray of light from a source 17.6 cm from a mirror strikes the mirror at the point shown. How far is the screen S from the mirror?

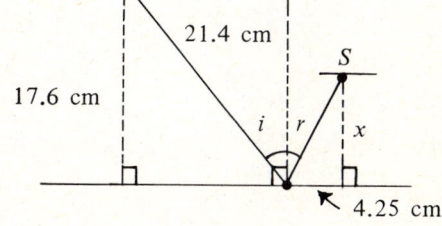

55. What is the length x of the steel support in the structure shown in the figure at the right?

56. Lots A and B extend from First Street to Main Street as shown in the figure at the right. Lot A has frontages of 200 ft and 140 ft on First Street and Main Street, respectively. Lot B has a frontage of 120 ft on First Street. What is the frontage of Lot B on Main Street?

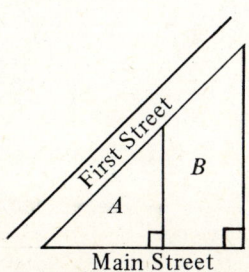

UNIT THREE: THE CIRCLE

3-1 BASIC DEFINITIONS AND PROPERTIES

3-1-1 We now consider another important geometric figure, the *circle*. There are many important applications associated with the circle in science and technology, including its use in architecture and modern machinery.

3-1-2 All points on a circle are the same distance from a fixed point in the plane. (See the figure below at the left.) The fixed point O is the *center* of the circle.

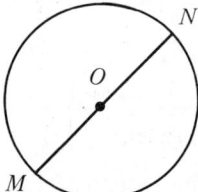 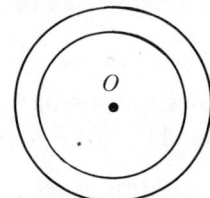

The distance ON (or OM) from the center to a point on the circle is the *radius* of the circle. The distance MN between two points on the circle and on a line passing through the center is the *diameter* of the circle. Thus, the diameter d is twice the radius r, or $d = 2r$. Two circles with the same center are *concentric*, as shown in the figure at the right.

3-1-3 **EXAMPLE 3-1-A**

In the circle shown, O is the center, PO and OQ are radii (plural of radius), and PQ is a diameter. If a radius such as OQ is 3 cm in length, then $r = 3$ cm and any diameter $d = 6$ cm.

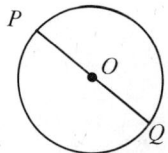

EXERCISES 3-1-A

In the given circle:
1. identify two radii,
2. identify a diameter.
3. If $SO = 2$ in., $ST = ?$
4. If $d = 5$ m, $r = ?$

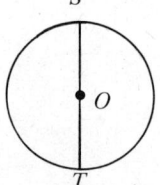

3-1-4   A line segment having its end points on the circle is a *chord*. A *tangent* is a line that touches a circle at one point (does not pass through). A *secant* is a line that passes through two points of a circle. These are illustrated in the following example.

3-1-5   **EXAMPLE 3-1-B**

For the circle shown at the right, line segment *AB* is called a chord, line *SU* is a secant line, and line *TV*, which passes through point *B*, is a tangent line. *CB* is a diameter and is also a chord.

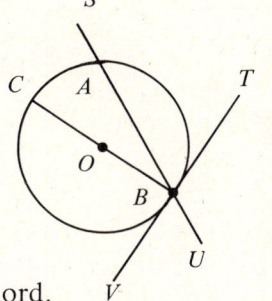

EXERCISES 3-1-B

For the given circle, identify:

1. Two chords
2. A secant line
3. A tangent line
4. A diameter

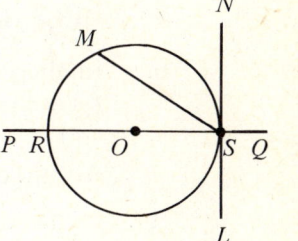

3-1-6   A *central angle* of a circle is an angle with vertex at the center of the circle. An *arc* of a circle consists of that part of the circle between and containing two specified points. There are two such arcs on a circle, the minor arc and the major arc. An arc is measured by its central angle. Central angles and arcs are illustrated in the following example.

3-1-7   **EXAMPLE 3-1-C**

For the circle shown at the right, ∠*DOC* is a central angle (we could also have designated it as ∠*AOB*). Also, that part of the circle between and including *A* and *B* is the arc *AB*. There are two arcs: the minor arc *AB*, designated as $\widehat{AB}$, and the major arc *AEB*, designated $\widehat{AEB}$. We shall use two-letter arc designations only for minor arcs. If ∠*DOC* = 50°, then $\widehat{AB}$ = 50°, and $\widehat{AEB}$ = 310°. We note that a minor arc has a measure less than 180° and a major arc has a measure between 180° and 360°.

EXERCISES 3-1-C

For the circle shown:

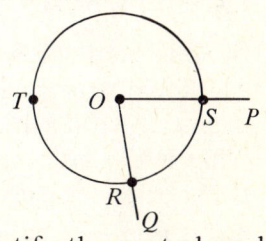

1. identify the central angle,
2. identify the minor arc.
3. If the measure of ∠*POQ* = 80°, find the measure of $\widehat{SR}$.
4. If ∠*POQ* = 80°, find the measure of $\widehat{STR}$.

56 The Circle

3-1-8   An angle is *inscribed* in an arc if the sides of the angle contain the end points of the arc, and the vertex of the angle is a point (not an end point) of the arc. An important property associated with inscribed angles is that *the measure of an inscribed angle is one-half of its intercepted arc.* This is illustrated in the following examples.

3-1-9   **EXAMPLE 3-1-D**

(a) In the circle at the right, $\angle ABC$ is inscribed in $\overset{\frown}{ABC}$, and it intercepts $\overset{\frown}{AC}$. If $\overset{\frown}{AC} = 60°$, $\angle ABC = 30°$.

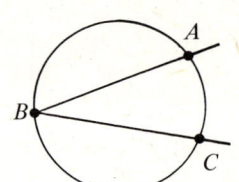

(b) In the circle at the right, PQ is a diameter and $\angle PRQ$ is inscribed in semi-circular $\overset{\frown}{PRQ}$. Since $\overset{\frown}{PSQ} = 180°$, $\angle PRQ = 90°$. We conclude that an angle inscribed in a semicircle is a right angle.

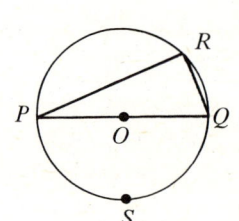

EXERCISES 3-1-D

For Exercises 1 and 2, use the left circle, and for Exercises 3 and 4, use the right circle.

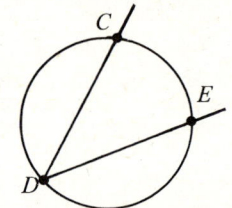 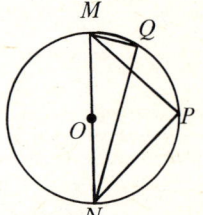

1. Identify the inscribed angle and the intercepted arc.
2. If $\overset{\frown}{CE} = 84°$, $\angle CDE = ?$
3. If MN is a diameter, $\angle MQN = ?$
4. If MN is a diameter, $\angle MPN = ?$

3-1-10 **EXAMPLE 3-1-E**

In constructing a wheel, a metal equilateral triangular support is placed within the rim as shown. What is the intercepted arc between support pieces AB and BC?

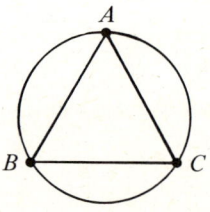

Since $\triangle ABC$ is equilateral, $\angle ABC = 60°$. Since the inscribed angle is one-half the intercepted arc, $\overset{\frown}{AC} = 120°$.

EXERCISES 3-1-E

In the figure for Example 3-1-E, find the indicated arcs if $AB = AC$ and $\angle BAC = 50°$.

1. $\overset{\frown}{BC}$
2. $\overset{\frown}{AC}$

3-1-11 An important property of a tangent line to a circle is that *a tangent to a circle is perpendicular to the radius drawn to the point of contact.* This is illustrated in the following example.

3-1-12   EXAMPLE 3-1-F

In the figure at the right, O is the center of the circle and PQ is tangent at Q. If ∠OPQ = 25°, find the measure of ∠POQ.

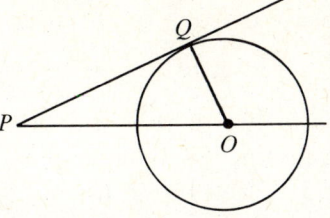

Since the center of the circle is point O, OQ is a radius. A tangent is perpendicular to a radius at the point of tangency, which means ∠OQP = 90°. Thus, ∠OPQ + ∠OQP = 25° + 90° = 115°. The sum of the angles of a triangle is 180°. Therefore, ∠POQ = 180° − 115° = 65°.

EXERCISES 3-1-F

In the figure below, O is the center of the circle, AB is tangent at B, and AC is tangent at C. ∠OAC = 30° and ∠AOC = ∠AOB.

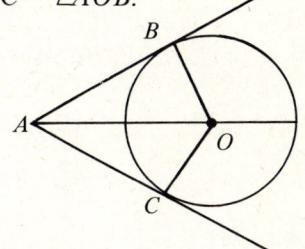

1. Find the measure of ∠AOB.
2. Find the measure of $\overset{\frown}{BC}$.

3-1-13   In this section we have introduced the circle and basic definitions and properties associated with it. We have introduced the center, radius, diameter, chord, tangent, secant, arc, central angle, and inscribed angle. Among the important properties we discussed were (1) an arc is measured by its central angle, (2) the measure of an inscribed angle is one-half of its intercepted arc, and (3) a tangent to a circle is perpendicular to the radius drawn to the point of contact. The following exercises provide an opportunity to review the definitions and properties of the circle.

3-1-14   EXERCISES 3-1-Section

In Exercises 1-12, use the figure at the right. In the figure, O is the center of the circle. Identify the following:

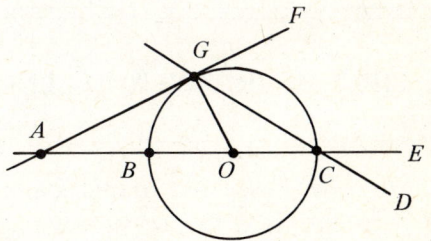

1. Three radii
2. A diameter
3. Two secant lines
4. A tangent line
5. Two chords
6. An isosceles triangle
7. Two minor arcs
8. Two major arcs
9. An acute central angle
10. An inscribed angle
11. Two perpendicular lines
12. If OG = 8 cm, BC = ?

58   The Circle

In Exercises 13-16, use the same figure as that used for Exercises 1-12. If
∠BOG = 60°, determine the measure of each of the following:

13. $\stackrel{\frown}{BG}$ 14. $\stackrel{\frown}{GC}$ 15. ∠BCG   16. ∠GAO

In Exercises 17-24, use the figure at the right. In the
figure, O is the center of the circle, line BT is tangent
to the circle at B, and ∠ABC = 55°. Determine the
measure of the following arcs and angles.

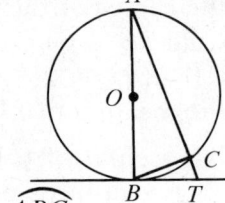

17. $\stackrel{\frown}{AC}$ 18. $\stackrel{\frown}{BC}$ 19. $\stackrel{\frown}{CAB}$ 20. $\stackrel{\frown}{ABC}$
21. ∠BCT   22. ∠CBT   23. ∠CAB   24. ∠BTC

In Exercises 25-28, solve the given problems.

25. The earth moves in an approximately circular orbit around the sun. The
diameter of the orbit is 186,000,000 mi. How far is the earth from the sun?

26. The chord between the ends of two radii of a circle is the same length as the
radius. What is the central angle between the radii?

27. A square support frame is constructed in a circular rim such that the vertices
of the square are in contact with the rim. What is the measure of the arc
intercepted by adjacent sides of the square?

28. The pulley belt shown is in contact with
pulley along an arc of 220° What is the
indicated angle between sections of the belt?

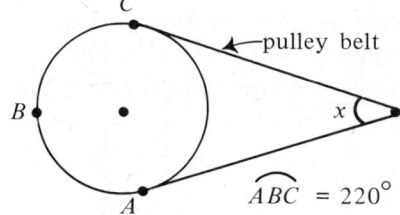

3-2   CIRCUMFERENCE AND LENGTH OF ARC

3-2-1   When we discussed triangles and quadrilaterals we introduced the concept of
perimeter. In this section we consider the perimeter and the length of an arc of
a circle, along with some of the elementary applications.

3-2 Circumference and Length of Arc

3-2-2 The distance around a circle, or its perimeter, is called the *circumference*. We cannot directly develop a formula for the circumference of a circle, as we did for the perimeters of triangles and quadrilaterals, for there are no sides to identify and measure. However, we will use a basic geometric fact, without attempting to develop it, in order to arrive at the necessary formulas. It is that *the ratio of the circumference to the diameter is the same for all circles*. This ratio is represented by π, a number equal to approximately 3.14. To six significant digits, $\pi = 3.14159$. We will use the decimal form, rather than the mixed number form $3\frac{1}{7}$ (which is also only approximate) for π, since the decimal form is generally more convenient for calculations.

3-2-3 Since the ratio of the circumference c to the diameter d equals π, we have $\frac{c}{d} = \pi$. This leads to the formula

$$c = \pi d$$

for the circumference. Also, since the diameter equals twice the radius, we have

$$c = 2\pi r$$

as a form of a formula for the circumference in terms of the radius.

3-2-4 **EXAMPLE 3-2-A**

(a) If the diameter of a circle is 3 in., we have

$$c = \pi(3) = 3\pi \text{ in.}$$

as the circumference. It is reasonably common to leave answers involving π in terms of π.

(b) The answer in terms of π is convenient, since the calculation is not required. However, it is not particularly useful from a point of view of measurement. Thus, using $\pi = 3.14$, we have

$$c = \pi(3) = 3\pi = 3(3.14) = 9.42 \text{ in.}$$

as the circumference of the circle in part (a).

(c) If the radius of a circle is 2.72 cm, the circumference is

$$c = 2\pi(2.72) = 2(3.14)(2.72) = 17.1 \text{ cm.}$$

The result has been rounded off to three significant digits.

EXERCISES 3-2-A

In Exercises 1 and 2, determine the circumferences of the circles (a) in terms of π, and (b) to three significant digits using $\pi = 3.14$ for the given values of d and r.

1. $d = 8$ ft
2. $r = 20$ m

In Exercises 3 and 4, determine the circumferences of the circles to three significant digits using $\pi = 3.14$ for the given values of d and r.

3. $d = 86.5$ mm
4. $r = 0.962$ in.

60 The Circle

3-2-5 The measure of the arc of an entire circle is 360°. The measure of an arc is the same as its central angle. Therefore, the ratio of the *length of arc* of a circle to the circumference of the circle is the same as the ratio of the measure of the central angle of the arc to 360°. This gives us

$$\frac{L}{2\pi r} = \frac{n}{360°},$$

where L is the length of arc and n is the measure in degrees of the central angle of the arc. This can be written as

$$L = \frac{n\pi r}{180°}.$$

3-2-6 **EXAMPLE 3-2-B**

(a) The length of arc L in the figure at the right is

$$L = \frac{120° \pi (2.50)}{180°}$$

$$= \frac{2(3.14)(2.50)}{3} = 5.23 \text{ in.}$$

(b) The length of arc L where $n = 80°$ and $r = 14.6$ cm is

$$L = \frac{80° \pi (14.6)}{180°} = \frac{4(3.14)(14.6)}{9}$$

$$= 20.4 \text{ cm}.$$

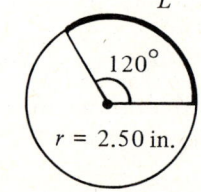

EXERCISES 3-2-B

Find the lengths of arc given the following radii and central angles.

1. 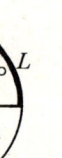   2.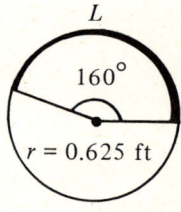

3. $r = 18.3$ in., $n = 50°$

4. $r = 495$ m, $n = 200°$

3-2-7 There are occasions when the circumference or length of arc is known and we are to find the radius of the circle. The following example illustrates the solution of this type of problem.

3-2 Circumference and Length of Arc

3-2-8 **EXAMPLE 3-2-C**

(a) The circumference of a circle is 32π cm long. To find the radius, we substitute 32π for c in the formula and then solve for r. This gives us

$$32\pi = 2\pi r$$
$$16 = r, \text{ or } r = 16 \text{ cm}.$$

(b) The length of arc, of which the central angle is $60°$, is 15.6 in. The radius is found as follows:

$$15.6 = \frac{60° \pi r}{180°},$$

$$15.6 = \frac{3.14 r}{3}$$

$$r = \frac{3(15.6)}{3.14} = 14.9 \text{ in}.$$

EXERCISES 3-2-C

In Exercises 1 and 2, determine the radius of the circle for which the circumference is given.

1. $c = 20\pi$ ft   2. $c = 36.0$ m

In Exercises 3 and 4, determine the radius of the circle for which the length of an arc and measure of its central angle are given.

3. $L = 10\pi$ mm, $n = 45°$

4. $L = 1.08$ mi, $n = 10°$

3-2-9 Circumference and arc length are important in certain applied situations. The following example illustrates an application involving an arc length.

3-2-10 **EXAMPLE 3-2-D**

A Norman window is one whose shape is a rectangle surmounted by a semicircle as shown in the figure at the right. Find the perimeter of a Norman window, given that its vertical side h is 1.20 m and the radius of the circular part is 0.550 m.

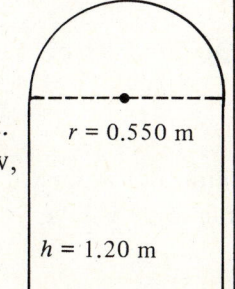

$r = 0.550$ m

$h = 1.20$ m

The perimeter is the sum of the lengths of the two vertical sides, the base (which equals the diameter of the circular part), and the semicircular part at the top. Thus,

$$p = 2h + 2r + \frac{180° \pi r}{180°} = 2h + 2r + \pi r$$

$$= 2(1.20) + 2(0.550) + (3.14)(0.550)$$

$$= 2.40 + 1.10 + 1.73 = 5.23 \text{ m}.$$

EXERCISES 3-2-D

1. Find the perimeter of a Norman window for which $h = 6.50$ ft and $r = 2.25$ ft.

2. A certain machine part is in the shape of a quarter circle of radius 17.3 mm. Find the perimeter of the part. See the figure below.

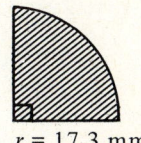

$r = 17.3$ mm

62   The Circle

3-2-11   In this section we have discussed the measurement of the circumference and length of arc of a circle. The following exercises provide an opportunity to review the calculation of these measures.

3-2-12   **EXERCISES 3-2-Section**

In all of the following exercises, express answers to three significant digits, unless they are expressed in terms of π. Use $\pi = 3.14$.

In Exercises 1-4, determine the circumferences of the circles (a) in terms of π, and (b) to three significant digits for the given values of d and r.

1. $d = 10$ ft   2. $d = 4$ m   3. $r = 5$ mm   4. $r = 15$ in.

In Exercises 5-12, determine the circumferences of the circles for the given values of d and r.

5. $d = 20.0$ in.   6. $d = 6.00$ mm   7. $d = 0.660$ m   8. $d = 520$ ft

9. $r = 86.3$ yd   10. $r = 37.1$ cm   11. $r = 2.92$ km   12. $r = 0.117$ mi

In Exercises 13-20, determine the lengths of arc with the given radii and central angles.

13. $r = 47.0$ cm, $n = 100°$ 14. $r = 7.30$ in., $n = 140°$

15. $r = 0.442$ ft, $n = 20°$ 16. $r = 285$ mm, $n = 42°$

17. $r = 17.2$ m, $n = 210°$ 18. $r = 55.0$ ft, $n = 240°$

19. $r = 61.5$ in., $n = 75°$ 20. $r = 3.09$ cm, $n = 35°$

In Exercises 21-24, determine the radius of the circle for which the circumference or length of arc and central angle is given.

21. $c = 48.2$ in.   22. $c = 4.95$ m

23. $L = 22.7$ cm, $n = 150°$ 24. $L = 215$ ft, $n = 15°$

In Exercises 25-32, solve the given problems.

25. The diameter of an automobile tire is 28.0 in. What is the circumference of the tire?

26. The inner diameter of a water pipe is 2.54 cm. What is the inner circumference of the pipe?

27. A cam is constructed such that part of it is a circular arc with a central angle of 72° and a radius of 5.30 mm. What is the length of arc along this part of the cam?

28. The minute hand of a clock is 4.75 in. long. Through what distance does the end of the minute hand move in 20 minutes?

29. To measure the diameter of a tree a rope is placed around it and then measured. If the length of rope placed around the tree is 6.50 ft, what is the diameter of the tree?

30. In traveling one-fourth of the way around a traffic circle a car travels 0.075 km. What is the radius of the traffic circle?

31. What is the perimeter of the area shown in the figure at the right? It is a quarter-circle-rectangle combination.

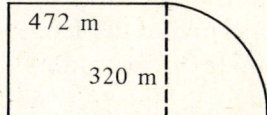

32. Find the length of the pulley belt shown in the figure at the right.

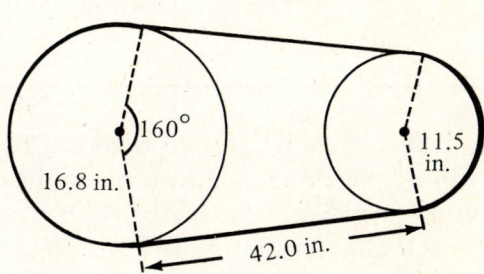

3-3   AREA

3-3-1   We now consider the area of a circle and a sector of a circle. Again, we cannot directly develop a formula for the area of a circle, but will use the formula which is found through more advanced methods.

3-3-2   As in the case of the circumference of a circle, the formula for the area of a circle is expressed in terms of π. The area of a circle is

$$A = \pi r^2.$$

64 The Circle

3-3-3 **EXAMPLE 3-3-A**

(a) The area of the circle at the right is

$$A = \pi(2.73)^2$$
$$= (3.14)(7.4529)$$
$$= 23.4 \text{ sq cm.}$$

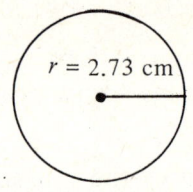
$r = 2.73$ cm

(b) To find the area of a circle of given diameter, we first divide the length of the diameter to obtain the length of the radius. Thus, if $d = 48.2$ in., $r = 24.1$ in. and

$$A = \pi(24.1)^2 = (3.14)(580.81)$$
$$= 1820 \text{ sq in.}$$

EXERCISES 3-3-A

In the following exercises, determine the areas of the circles to three significant digits using $\pi = 3.14$ for the given values of d and r.

1. $r = 10.0$ ft
2. $r = 58.0$ m
3. $d = 184$ mm
4. $d = 0.728$ mi

3-3-4 A *sector* of a circle is a region bounded by an arc and two radii of the circle as shown in the figure at the right. The ratio of the area of a sector to that of the entire circle is the same as the ratio of the measure of the central angle of the arc to 360°. This gives us

$$\frac{A}{\pi r^2} = \frac{n}{360°},$$

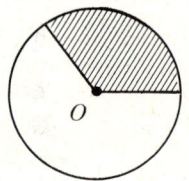

where A is the area of the sector and n is the measure in degrees of the central angle of the arc. Thus,

$$A = \frac{n\pi r^2}{360°}.$$

3-3-5 **EXAMPLE 3-3-B**

(a) The area of the sector shown in the figure at the right is

$$A = \frac{70°\pi(16.3)^2}{360°}$$
$$= \frac{7(3.14)(265.69)}{36}$$
$$= 162 \text{ sq cm.}$$

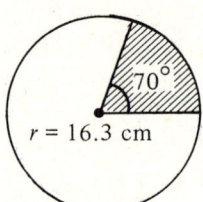
$r = 16.3$ cm

(b) The area of the sector of a circle for which $n = 115°$ and $r = 2.80$ ft is

$$A = \frac{115°\pi(2.80)^2}{360°} = \frac{23(3.14)(7.84)}{72} = 7.86 \text{ sq ft.}$$

EXERCISES 3-3-B

Find the areas of the sectors with the given radii and central angles.

1.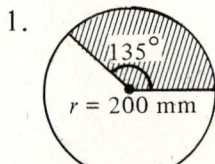
$r = 200$ mm

2.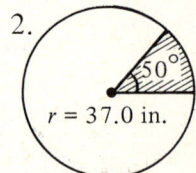
$r = 37.0$ in.

3. (quarter circle)
$n = 90°$, $r = 75.0$ ft

4. $n = 150°$, $r = 2.91$ m

3-3-6   The following example illustrates the type of problem in which the area, or the area of a sector, is known and we are to solve for the radius of the circle.

3-3-7   **EXAMPLE 3-3-C**

(a) The area of a circle is 64π sq m. To find the radius we substitute 64π for A in the formula and then solve for r. This gives us

$$64\pi = \pi r^2,$$
$$r^2 = 64.$$

Since $r^2 = 64$, $r = \sqrt{64}$ or $r = 8$ m.

(b) The area of a circular sector, of which the central angle is $75°$, is 22.9 sq ft. Thus, to find r, we have

$$22.9 = \frac{75° \pi r^2}{360°}, \quad 22.9 = \frac{5(3.14)r^2}{24}$$

$$r^2 = \frac{(24)(22.9)}{5(3.14)} = 35.01; \quad r = 5.92 \text{ ft.}$$

EXERCISES 3-3-C

In Exercises 1 and 2, determine the radius of the circle for which the area is given.

1. $A = 25\pi$ sq cm

2. $A = 2.92$ sq in.

In Exercises 3 and 4, determine the radius of the circle for which the area of a sector and measure of its central angle are given.

3. $A = 54.0$ sq ft, $n = 60°$

4. $A = 822$ sq mm, $n = 105°$

3-3-8   The following example illustrates the use of the area of a circular sector in an applied situation.

3-3-9   **EXAMPLE 3-3-D**

A patio is in the shape of a rectangle, with a quarter circle removed, as shown in the figure at the right. What is the area of the patio?

To find the area of the patio we must subtract the quarter-circular area from the rectangular area. Thus, we have

$$A = (65.0)(17.5 + 12.5) - \frac{90° \pi (17.5)^2}{360°}$$

$$= (65.0)(30.0) - \frac{(3.14)(17.5)^2}{4}$$

$$= 1950 - 240 = 1710 \text{ sq ft.}$$

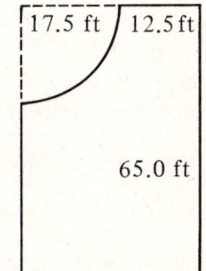

17.5 ft   12.5 ft

65.0 ft

EXERCISES 3-3-D

1. What is the area of the Norman window of Example 3-2-D?

2. A flat steel ring 17.2 in. in diameter has a hole 3.20 in. in diameter in the center. What is the area of one face of the ring?

3-3-10   In this section we have discussed the measurement of the area of a circle and the area of a circular sector. The following exercises provide an opportunity to review the calculation of these measures.

3-3-11   **EXERCISES 3-3-Section**

In all of the following exercises, express answers to three significant digits. Use $\pi = 3.14$.

In Exercises 1-8, determine the areas of the circles for the given values of r and d.

1. $r = 40.0$ in.   2. $r = 8.00$ mm   3. $r = 0.940$ m   4. $r = 270$ ft
5. $d = 78.2$ yd   6. $d = 7.46$ km   7. $d = 92.8$ cm   8. $d = 0.404$ mi

In Exercises 9-16, determine the areas of the sectors with the given radii and central angles.

9. $r = 72.0$ cm, $n = 110°$ 10. $r = 8.90$ in., $n = 160°$
11. $r = 0.666$ ft, $n = 40°$ 12. $r = 107$ mm, $n = 48°$
13. $r = 19.2$ m, $n = 225°$ 14. $r = 8.61$ ft, $n = 250°$
15. $r = 78.5$ in., $n = 36°$ 16. $r = 0.929$ cm, $n = 72°$

In Exercises 17-20, determine the radius of the circle for which the area or the area of a sector and its central angle is given.

17. $A = 92.0$ sq in.   18. $A = 9.30$ sq m
19. $A = 604$ sq cm, $n = 120°$ 20. $A = 67.3$ sq ft, $n = 108°$

In Exercises 21-28, solve the given problems.

21. What is the cross-sectional area of a pipe whose diameter is 5.08 cm?

22. What is the cross-sectional area of a wire whose diameter is 1.50 mm?

23. A flower bed is in the shape of a semicircle of radius 12.0 m. What is the area of the flower bed?

24. The bottom of a storage tank has the shape of the circular sector shown in the figure at the right. What is the total force on the bottom of the tank if the pressure is 7250 pounds per square foot?

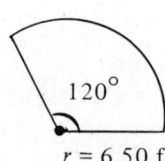

25. What should be the diameter of an exhaust pipe if its area is 12.0 sq ft?

26. A hole of area 5.25 sq cm is to be drilled in a piece of sheet metal. What should be the radius of the hole?

27. A circular walk 2.25 m wide is to be built around a circular pool 19.5 m in diameter. What is the area of the walk?

28. A race track is constructed in the shape of a rectangle with semicircles at each end. The lengths of the semicircular arcs and the straight sections are each 0.250 mi. What is the area within the track? See the figure at the right.

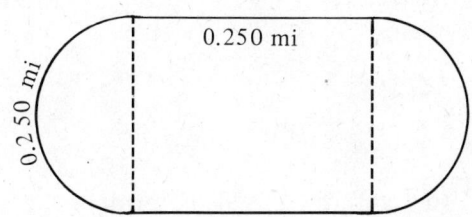

3-4 EXERCISES FOR UNIT THREE

In Exercises 1-4, use the figure at the right. Line AT is tangent to the circle with center at O. Determine the measures of the indicated arcs and angles.

1. $\overparen{BT}$
2. $\overparen{BCT}$
3. $\angle OTA$
4. $\angle TAO$

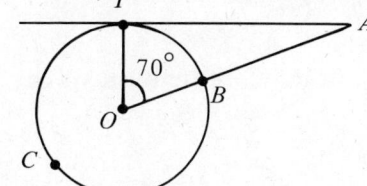

In Exercises 5-8, use the figure at the right. Determine the measures of the indicated arcs and angles.

5. $\overparen{BC}$
6. $\overparen{AB}$
7. $\angle ABC$
8. $\angle ACB$

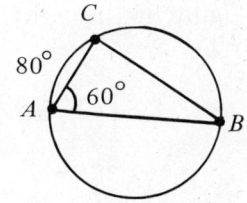

In Exercises 9-12, use the figure at the right. Line CT is tangent to the circle with center at O. Determine the measures of the indicated angles.

9. $\angle BTA$
10. $\angle TAB$
11. $\angle BTC$
12. $\angle ABT$

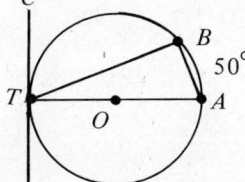

In the following exercises, express answers to three significant digits. Use $\pi = 3.14$.

In Exercises 13-16, determine the circumferences of the circles for the given values of d and r.

13. $d = 17.0$ cm   14. $d = 38.0$ in.   15. $r = 4.25$ ft   16. $r = 8.58$ m

In Exercises 17-20, determine the lengths of arc with the given radii and central angles.

17. $r = 52.5$ in., $n = 18°$ 18. $r = 22.7$ cm, $n = 54°$

19. $r = 320$ mm, $n = 96°$ 20. $r = 1.67$ mi, $n = 144°$

In Exercises 21-24, determine the areas of the circles for the given values of r and d.

21. $r = 3.36$ cm   22. $r = 81.2$ in.   23. $d = 0.608$ ft   24. $d = 420$ mm

In Exercises 25-28, determine the areas of the sectors with the given radii and central angles.

25. $r = 40.5$ in., $n = 130°$ 26. $r = 72.8$ cm, $n = 110°$

27. $r = 3.06$ m, $n = 66°$ 28. $r = 7.11$ ft, $n = 6°$

In Exercises 29-32, determine the radius of the circle for which the circumference or area is given.

29. $c = 465$ mm   30. $c = 2.33$ in.

31. $A = 1730$ sq ft   32. $A = 56,400$ sq m

In Exercises 33-34, use the figure at the right. The figure is a combination of a right triangle and a semicircle.

33. Find the perimeter.

34. Find the area.

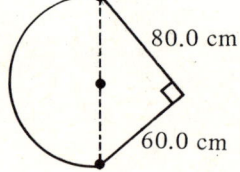

In Exercises 35-36, use the figure at the right. The figure is composed of two circular sectors.

35. Find the area.

36. Find the perimeter.

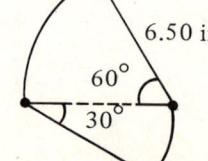

In Exercises 37-44, solve the given problems.

37. The radius of the earth is 3960 mi. An artificial satellite is traveling around the earth in a circular orbit at an altitude of 180 mi. What is the circumference of the earth and how far does the satellite travel in one orbit of the earth?

38. An ammeter needle is 7.50 cm long. Through what distance does the end of the needle move when it is deflected through an angle of 135°?

39. A flexible molding is to be placed along the sides and top of the entrance way shown in the figure at the right. The entrance way is rectangular with a semicircular top. What length of molding is needed?

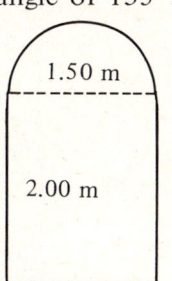

40. To help resist wear, the edge of the cam in the figure at the right is to be coated with a special metal strip costing $1.25 per inch. How much will it cost to put the strip on the cam? The figure consists of a semicircle, two quarter-circles, and a rectangle.

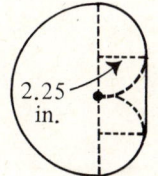

41. The diameter of a piston is 24.0 in. Find the total force on the piston when the pressure is 150 pounds per square inch.

42. How many pipes 2.00 cm in radius are required to carry as much water as a pipe 6.00 cm in radius?

43. What is the area of the hallway shown in the figure at the right? It consists of two rectangles and a quarter circle.

44. The arm of a car windshield wiper is 12.5 in. long and is attached at the middle of a 15.0-in. blade. (Assume the arm and blade are in line.) What area of the windshield is cleaned by the wiper if it swings through 100° arcs?

UNIT FOUR: SOLID GEOMETRIC FIGURES

4-1 PRISMS

4-1-1 We are now going to consider some of the measurements associated with three-dimensional geometric figures. These figures are used in architecture and engineering, as well as other areas of technology. In our discussion of solid figures in this unit, we will consider volumes and areas of these figures.

4-1-2 A *polyhedron* is a geometric solid figure which is bounded by planes. The plane surfaces of the polyhedron are called *faces*, and the intersections of the faces are called *edges*. In the figure at the right, a polyhedron of four faces and six edges is illustrated.

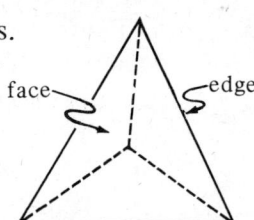

4-1-3 A *prism* is a polyhedron whose *bases* are parallel and equal polygons and whose side faces are parallelograms. In the figure at the right, a prism with triangular bases is shown. We will consider only prisms in which the side edges are perpendicular to the bases (right prisms).

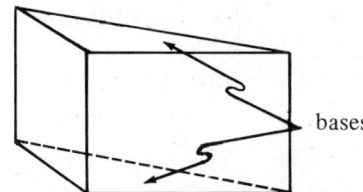

4-1-4 A prism of particular importance is a *rectangular solid*. In a rectangular solid, all six faces are rectangles. A rectangular solid is shown in the figure at the right.

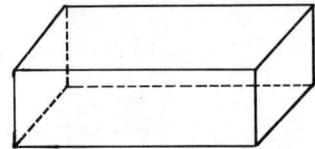

4-1-5 An important measure associated with a solid figure is its *volume*. Just as area gives us a measure of the surface of a geometric figure, volume gives us a measure of the amount of space occupied by the figure. In finding the volume of a solid geometric figure we are finding the number of cubes, one unit on an edge, which are required to fill the figure. In the rectangular solid at the right, we see that 24 cubes, each 1 cm on an edge, are required to fill the solid. Thus, we can say that the volume of the solid is 24 cu cm.

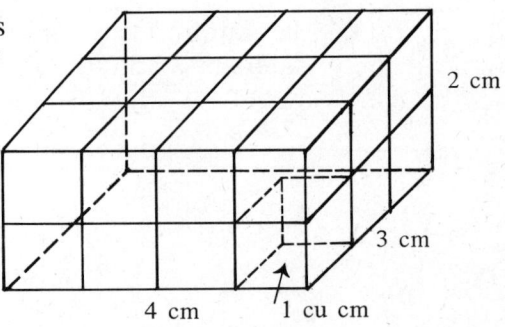

4-1-6   If we note the rectangular solid in Article 4-1-5, we see that its volume may be determined by finding the product of its length, width, and height. Thus, the volume of any rectangular solid is

$$V = lwh,$$

where l, w, and h are in the same unit of length, and V is in cubic units of length. Where appropriate, we shall round off all calculations to three significant digits.

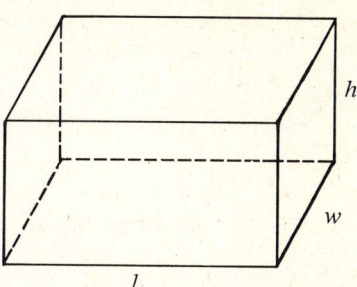

4-1-7   **EXAMPLE 4-1-A**

(a) The volume of a rectangular solid of length 6 in., width 4 in., and height 3 in., is

$$V = (6)(4)(3) = 72 \text{ cu in.}$$

(b) The volume of a rectangular solid for which $l = 2.75$ m, $w = 1.83$ m, and $h = 1.25$ m is

$$V = (2.75)(1.83)(1.25) = 6.29 \text{ cu m.}$$

EXERCISES 4-1-A

Find the volumes of the rectangular solids with the given lengths, widths, and heights.

1. $l = 5$ ft, $w = 2$ ft, $h = 8$ ft
2. $l = 10$ cm, $w = 6$ cm, $h = 8$ cm
3. $l = 14.5$ mm, $w = 10.3$ mm, $h = 11.6$ mm
4. $l = 26.5$ in., $w = 23.2$ in., $h = 15.6$ in.

4-1-8   The *cube*, which we have already used in naming units of volume, is a prism in which all six faces are squares and in which all edges are equal. Therefore, since the length, width, and height equal the length of an edge e, the volume of a cube is

$$V = e^3.$$

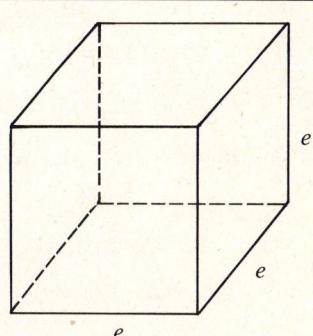

4-1-9   **EXAMPLE 4-1-B**

(a) The volume of a cube of edge 7 in. is

$$V = 7^3 = 343 \text{ cu in.}$$

(b) The volume of a cube for which $e = 0.936$ m is

$$V = (0.936)^3 = 0.820 \text{ cu m.}$$

EXERCISES 4-1-B

Find the volumes of the cubes with the given edges.

1. $e = 4$ ft
2. $e = 10$ cm
3. $e = 89.3$ mm
4. $e = 16.2$ yd

72 Solid Geometric Figures

4-1-10 The base of a rectangular solid is a rectangle of length l and width w. The area of this rectangle is lw. Calling B the area of this base we can express the formula for the volume as $V = Bh$. Since we can find the volume knowing B and h, we may conclude that the volume of any prism is

$$V = Bh,$$

where B is the area of the base and h is the height, or altitude, of the prism.

4-1-11 **EXAMPLE 4-1-C**

(a) The base of a prism is a quadrilateral of area 75.0 sq cm. The height of the prism is 6.51 cm. The volume of the prism is
$V = (75.0)(6.51)$
$ = 488$ cu cm.

(b) The base of a prism is a right triangle with legs of 14.6 in. and 10.9 in. The height of the prism is 12.3 in. To find the volume we first find the area of the base to be one-half the product of the legs of the triangle. Thus,

$B = \frac{1}{2}(14.6)(10.9)$
$ = 79.6$ sq in.

This means the volume is

$V = (79.6)(12.3)$
$ = 979$ cu in.

EXERCISES 4-1-C

In Exercises 1 and 2, find the volumes of the prisms with given base areas and heights.

1. $B = 320$ sq in., $h = 14.0$ in.
2. $B = 4900$ sq m, $h = 72.0$ m

In Exercises 3 and 4, find the volumes of the prisms with right triangular bases, with the given legs, and given heights.

3. legs: 46.0 mm, 26.0 mm; $h = 15.4$ mm

4. legs: 6.25 ft, 8.34 ft; $h = 3.19$ ft

4-1-12 The *lateral area* L of a prism is the area of the faces other than the two bases. These faces are called the *lateral faces*. Thus, the lateral area can be found by multiplying the perimeter of a base by the height of the prism.

4-1-13 **EXAMPLE 4-1-D**

A rectangular solid is 20.3 cm long, 11.8 cm wide, and 13.1 cm high. The lateral area is the area of the front face, back face, and end faces combined. This area is

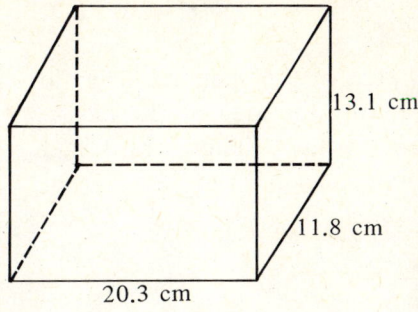

$$L = 2(20.3)(13.1) + 2(11.8)(13.1)$$
$$= 532 + 309 = 841 \text{ sq cm.}$$

We can also find the lateral area by first finding the perimeter of a base and multiplying by L. Thus,

$$p = 2(20.3) + 2(11.8) = 64.2 \text{ cm,}$$
$$L = (64.2)(13.1) = 841 \text{ sq cm.}$$

We see that the results agree.

EXERCISES 4-1-D

In Exercises 1 and 2, find the lateral areas of the rectangular solids for which the lengths, widths, and heights are given.

1. $l = 20$ in., $w = 15$ in., $h = 10$ in.
2. $l = 260$ mm, $w = 130$ mm, $h = 150$ mm

In Exercises 3 and 4, find the lateral areas of the prisms for the given bases and heights.

3. Base is triangle with sides 2.73 m, 3.20 m, 3.28 m; $h = 2.35$ m.
4. Base is quadrilateral with sides 17.0 ft, 16.5 ft, 23.2 ft, 32.5 ft; $h = 12.2$ ft.

4-1-14 The *total area* of a prism is the lateral area plus the area of the bases. In the case of a rectangular solid, there are three pairs of equal rectangular faces. Thus, a formula for the total area of a rectangular solid is

$$A = 2lw + 2wh + 2lh.$$

4-1-15 For a cube, each of the six faces is a square. Thus, the total area of a cube is

$$A = 6e^2.$$

4-1-16 EXAMPLE 4-1-E

(a) We may find the total area of the rectangular solid of Example 4-1-D by adding the area of the bases to the lateral area already found. The area of each base is

$$B = (20.3)(11.8) = 239.5 \text{ sq cm}.$$

Thus, the total area is

$$A = 2(239.5) + 841 = 479 + 841 = 1320 \text{ sq cm}.$$

(b) Using the formula for the area of a rectangular solid, we have

$$A = 2(20.3)(11.8) + 2(11.8)(13.1) + 2(20.3)(13.1)$$
$$= 479 + 309 + 532$$
$$= 1320 \text{ sq cm}.$$

(c) The area of a cube 6.82 in. on an edge is

$$A = 6(6.82)^2 = 6(46.5) = 279 \text{ sq in}.$$

EXERCISES 4-1-E

In Exercises 1 and 2, find the total areas of the rectangular solids for which the lengths, widths, and heights are given.

1. $l = 20$ in., $w = 15$ in., $h = 10$ in.
2. $l = 260$ mm, $w = 130$ mm, $h = 150$ mm

In Exercises 3 and 4, find the total areas of the cubes for which the length of an edge is given.

3. $e = 2.70$ m
4. $e = 17.8$ ft

4-1-17

In this section we introduced solid geometric figures and we discussed the volume, lateral area, and total area of prisms. The rectangular solid and cube are prisms of particular importance. The exercises which follow provide an opportunity to review the calculation of these measures of prisms, along with some of their applications.

4-1-18 EXERCISES 4-1-Section

In Exercises 1-16, determine the volumes of the prisms with the given dimensions.

1. Rectangular solid: $l = 6$ ft, $w = 5$ ft, $h = 4$ ft
2. Rectangular solid: $l = 30$ cm, $w = 25$ cm, $h = 40$ cm
3. Rectangular solid: $l = 82.0$ mm, $w = 17.0$ mm, $h = 18.0$ mm
4. Rectangular solid: $l = 2.55$ ft, $w = 2.05$ ft, $h = 1.20$ ft
5. Cube: $e = 16.0$ in.
6. Cube: $e = 0.900$ m
7. Cube: $e = 56.3$ cm
8. Cube: $e = 3.50$ yd
9. Triangular base: $B = 460$ sq ft, $h = 15.0$ ft

10. Triangular base: $B = 37.8$ sq m, $h = 5.27$ m
11. Quadrilateral base: $B = 2750$ sq cm, $h = 36.0$ cm
12. Polygon base: $B = 8630$ sq in., $h = 40.2$ in.
13. Base is right triangle with legs 2.37 ft and 4.22 ft; $h = 1.77$ ft.
14. Base is right triangle with legs 36.3 mm and 58.2 mm; $h = 12.8$ mm.
15. Base is parallelogram, for which base is 53.7 cm, height is 17.7 cm; $h = 23.5$ cm.
16. Base is parallelogram, for which base is 1.37 yd, height is 1.13 yd; $h = 1.27$ yd.

In Exercises 17-28, determine (a) the lateral areas and (b) the total areas of the prisms with given dimensions.

17. Rectangular solid: $l = 9$ cm, $w = 5$ cm, $h = 4$ cm
18. Rectangular solid: $l = 50.0$ in., $w = 20.0$ in., $h = 25.0$ in.
19. Rectangular solid: $l = 16.3$ ft, $w = 11.5$ ft, $h = 14.3$ ft
20. Rectangular solid: $l = 72.0$ mm, $w = 32.3$ mm, $h = 10.3$ mm
21. Cube: $e = 0.820$ m
22. Cube: $e = 2.73$ ft
23. Cube: $e = 32.7$ in.
24. Cube: $e = 15.2$ cm
25. Triangular base for which $B = 780$ sq mm, $p = 62.0$ mm; $h = 17.1$ mm
26. Triangular base for which $B = 17.2$ sq yd, $p = 16.3$ yd; $h = 3.37$ yd
27. Quadrilateral base for which $B = 3270$ sq in., $p = 238$ in.; $h = 51.8$ in.
28. Polygon base for which $B = 7360$ sq cm, $p = 421$ cm; $h = 42.9$ cm

In Exercises 29-36, solve the given problems.

29. What is the volume of air in a room 24.5 ft long, 15.0 ft wide, and 8.00 ft high?
30. How many cubic meters of concrete are needed for a driveway 20.0 m long, 2.75 m wide, and 0.100 m thick?
31. What is the weight of an ice cube 2.50 cm on an edge, if the density of ice is 0.917 grams per cubic centimeter?
32. A swimming pool is 1.10 m deep at one end and slopes uniformly to a depth of 3.40 m at the other end. The pool is 15.0 m long and 7.50 m wide. What is the volume of the pool? (The pool is a prism whose bases are its trapezoidal sides.)

33. A vertical beam has a rectangular cross-section 15.4 cm by 10.5 cm. The beam is 240 cm high. What is the lateral area of the beam?

34. A fish tank is 3.50 ft long, 2.75 ft wide, and 2.25 ft high. What is the surface area of the tank (there is no top)?

35. A rectangular box is to be used to store radioactive materials. The inside of the box is 12.0 in. long, 9.00 in. wide, and 9.00 in. deep. What is the area of sheet lead which must be used to line the inside of the box?

36. A glass prism used in the study of optics has a right triangular base. The legs of the right triangle are 3.00 cm and 4.00 cm. The prism is 8.50 cm high. What is the total surface area of the prism?

4-2 CYLINDERS

4-2-1 The base of a prism is a polygon. Another important solid figure, the *cylinder*, also has two equal bases, each of which is a curved figure. Cylinders are commonly found in pipes, containers, machinery, and in architectural structures.

4-2-2 In general, a *cylindrical surface* is *generated* by a straight line which always moves parallel to itself. A *cylinder* is a geometric solid bounded by a closed cylindrical surface and two parallel and equal bases. Thus, the bases of a cylinder may be circles, or other curved figures, and are not necessarily perpendicular to the cylindrical surface. The figures below illustrate various types of cylinders.

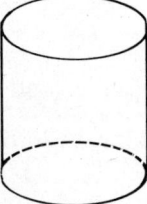  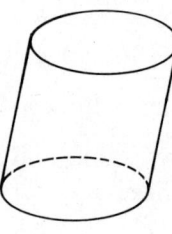

4-2-3   Although many types of cylinders may possibly be generated, one type is of particular importance. It is the *right circular cylinder*, which is generated by revolving a rectangle about one of its sides. In a right circular cylinder each base is a circle, and the cylindrical surface is perpendicular to the bases. We shall restrict our attention from this point to right circular cylinders, and whenever we use "cylinder" we shall mean "right circular cylinder."

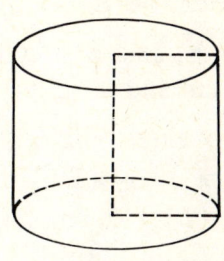

4-2-4   As with the prism, the volume of a right circular cylinder is the product of the area of a base and the height, or altitude, of the cylinder. Since the base is a circle, the area of a base is πr^2, and we have the formula

$$V = \pi r^2 h$$

for the volume, where r is the radius of a base and h is the height.

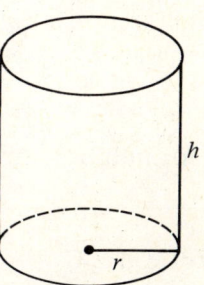

4-2-5	**EXAMPLE 4-2-A**	**EXERCISES 4-2-A**
	The volume of a right circular cylinder for which r = 8.25 in. and h = 6.75 in. is $V = \pi(8.25)^2(6.75)$ $ = (3.14)(68.0625)(6.75)$ $ = 1440$ cu in.	Determine the volumes of the cylinders with the given values of r and h. 1. r = 4.00 cm, h = 5.00 cm 2. r = 1.50 ft, h = 2.50 ft 3. r = 20.5 in., h = 15.0 in. 4. r = 0.375 m, h = 0.685 m

4-2-6   The lateral area of a prism is the product of the perimeter of a base and the height. In a similar manner, therefore, the lateral area of a cylinder is the circumference of a base times the height. We can think of this as the label on a can being removed and flattened out as demonstrated in the figure. Since the circumference of a base is $2\pi r$, the lateral area is

$$L = 2\pi r h.$$

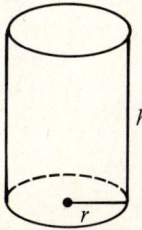 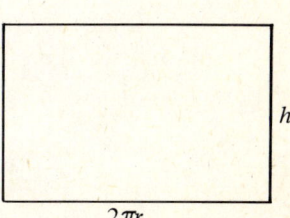

78   Solid Geometric Figures

4-2-7   **EXAMPLE 4-2-B**

The lateral area of a cylinder for which r = 3.50 cm and h = 2.60 cm is

$$L = 2\pi(3.50)(2.60)$$
$$= 2(3.14)(3.50)(2.60)$$
$$= 57.1 \text{ sq cm.}$$

EXERCISES 4-2-B

Determine the lateral areas of the cylinders with the given values of r and h.

1. r = 6.00 in., h = 5.00 in.
2. r = 16.0 m, h = 12.0 m
3. r = 25.5 mm, h = 46.3 mm
4. r = 2.33 ft, h = 6.28 ft

4-2-8   The total area of a right circular cylinder is the sum of the areas of the bases and the lateral area. Since each base is a circle of area πr^2, we have

$$A = 2\pi r^2 + 2\pi rh$$

as the formula for the total area.

4-2-9   **EXAMPLE 4-2-C**

The total area of a cylinder for which r = 3.50 cm and h = 2.60 cm (the values used in Example 4-2-B) is

$$A = 2\pi(3.50)^2 + 2\pi(3.50)(2.60)$$
$$= 2(3.14)(12.25) + 2(3.14)(3.50)(2.60)$$
$$= 76.9 + 57.1 = 134 \text{ sq cm.}$$

EXERCISES 4-2-C

Determine the total areas of the cylinders with the given values of r and h (the same as in Exercises 4-2-B).

1. r = 6.00 in., h = 5.00 in.
2. r = 16.0 m, h = 12.0 m
3. r = 25.5 mm, h = 46.3 mm
4. r = 2.33 ft, h = 6.28 ft

4-2-10 A common situation which arises in applications is the determination of the volume of a hollow cylinder as shown in the figure at the right. To do this we compute the volume of the outer cylinder and then subtract the volume of the inner hollow cylinder. Using R for the outer radius and r for the inner radius, we have

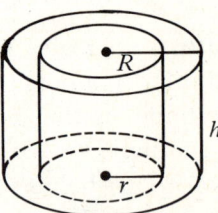

$$V = \pi R^2 h - \pi r^2 h$$

or

$$V = \pi h(R^2 - r^2).$$

4-2-11 **EXAMPLE 4-2-D**

The density of copper is 0.320 pounds per cubic inch. Find the weight of a piece of copper pipe 12.0 in. long with an inner radius of 0.500 in. and an outer radius of 0.625 in.

To find the weight we must find the volume of the copper in the pipe and then multiply by the density. From the given information we have $h = 12.0$ in., $r = 0.500$ in., and $R = 0.625$ in. Thus,

$$V = \pi(12.0)(0.625^2 - 0.500^2)$$
$$= (3.14)(12.0)(0.390625 - 0.250000)$$
$$= (3.14)(12.0)(0.140625)$$
$$= 5.30 \text{ cu in.}$$

Thus, the weight w is

$$w = (0.320)(5.30) = 1.70 \text{ lb.}$$

EXERCISES 4-2-D

1. Determine the weight of a piece of copper pipe 24.0 in. long with an inner radius of 1.00 in. and an outer radius of 1.25 in.

2. A layer of padding 1.20 cm thick is placed around a cylindrical bar which is 20.0 cm long and has a radius of 4.50 cm. What volume of padding is used?

4-2-12 In this section we introduced cylinders and we discussed the volume, lateral area, and total area of right circular cylinders. Also, we discussed the volume of the hollow cylinder. The exercises which follow provide an opportunity to review the calculations of these measures of cylinders.

4-2-13 **EXERCISES 4-2-Section**

In Exercises 1-8, determine the volumes of the cylinders with the given values of r, d (diameter of base), and h.

1. $r = 20.0$ cm, $h = 15.0$ cm
2. $r = 7.00$ in., $h = 4.00$ in.
3. $r = 15.0$ ft, $h = 3.60$ ft
4. $r = 1.58$ m, $h = 8.48$ m
5. $d = 366$ mm, $h = 140$ mm
6. $d = 0.634$ yd, $h = 0.156$ yd
7. $d = 24.2$ in., $h = 32.3$ in.
8. $d = 22.0$ cm, $h = 12.2$ cm

In Exercises 9-16, determine (a) the lateral area and (b) the total area of the cylinders with the given values of r, d, and h.

9. $r = 300$ mm, $h = 120$ mm
10. $r = 60.0$ ft, $h = 80.0$ ft
11. $r = 8.20$ in., $h = 2.40$ in.
12. $r = 2.30$ m, $h = 1.10$ m

13. $d = 24.0$ cm, $h = 8.50$ cm

14. $d = 3.80$ in., $h = 7.50$ in.

15. $d = 86.4$ ft, $h = 12.4$ ft

16. $d = 84.2$ mm, $h = 123$ mm

In Exercises 17-20, determine the volumes of the hollow cylinders with the given values of r, R, and h.

17. $r = 16.0$ in., $R = 20.0$ in., $h = 12.0$ in.

18. $r = 2.50$ cm, $R = 4.30$ cm, $h = 6.50$ cm

19. $r = 82.0$ mm, $R = 97.0$ mm, $h = 68.3$ mm

20. $r = 0.363$ ft, $R = 0.885$ ft, $h = 1.25$ ft

In Exercises 21-28 solve the given problems involving cylinders.

21. A cylindrical grain storage container 82.0 ft high has a radius of 24.3 ft. One bushel of grain occupies about 1.24 cu ft. How many bushels can be stored in the container?

22. An oil can has a radius of 12.5 cm and is 30.0 cm high. What is its volume in cubic meters? (1 cu m = 1,000,000 cu cm.)

23. What is the weight of one kilometer of copper wire which is 0.500 cm in diameter? The density of copper is 8.90 grams per cubic centimeter. (1 km = 100,000 cm.)

24. How many feet of steel rods 0.500 in. in radius can be made from a steel ingot which is in the shape of a rectangular solid 6.00 in. by 6.00 in. by 24.0 in.?

25. What is the area of a label (covering the lateral surface) on a cylindrical can 3.00 in. in diameter and 4.25 in. high? Assume the ends just meet and do not overlap.

26. A lawn roller is a cylinder 96.0 cm long and 30.0 cm in radius. What area is rolled in one complete revolution of the roller?

27. A hollow circular concrete conduit is 12.0 m long. The inside diameter is 54.0 cm and the outside diameter is 75.0 cm. Find the volume of concrete in the conduit.

28. A cylindrical water storage tank has an inside diameter of 24.0 ft and a height of 8.50 ft. It is made of material 6.00 in. thick. What volume of material is needed to construct the vertical cylindrical part of the tank?

4-3 PYRAMIDS AND CONES

4-3-1 The geometric solids we discussed in the previous sections of this unit have two parallel and equal bases. In this section we discuss solids with one base, and whose lateral surfaces meet at a vertex. Applications of these figures are found in architecture and engineering.

4-3-2 The base of a *pyramid* is a polygon. The other faces, the *lateral faces*, are triangles which meet at a common point, the *vertex*. The edges where the lateral faces meet are the *lateral edges*. The *height*, or *altitude*, of the prism is the perpendicular distance from the vertex to the base. These are illustrated in the figure at the right. We will consider only *regular pyramids*, those whose bases are regular polygons (equal sides and equal angles) and whose lateral faces are congruent (see Article 2-4-5) isosceles triangles.

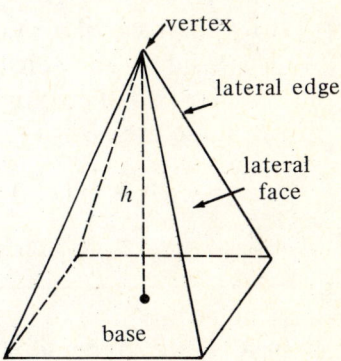

4-3-3 If we compare the volumes of a prism and a pyramid of the same base and height, we can see that the volume of the pyramid is considerably less than that of the prism. Although we cannot prove it here, the volume of the pyramid is exactly one-third of the volume of the prism. Thus,

$$V = \frac{1}{3}Bh$$

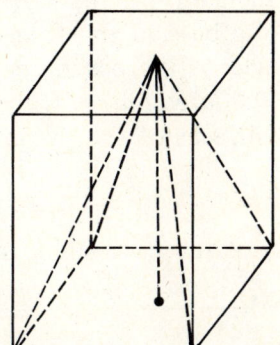
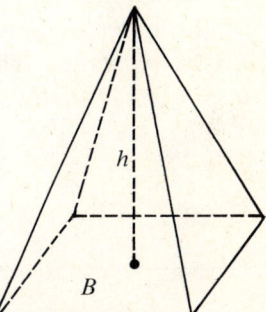

is the formula for the volume of a pyramid of base area B and height h.

4-3-4 EXAMPLE 4-3-A

(a) The base of a pyramid is a polygon of area 50.0 sq ft. The altitude of the pyramid is 15.0 ft. The volume of the pyramid is

$$V = \frac{1}{3}(50.0)(15.0)$$
$$= 250 \text{ cu ft.}$$

(b) The base of a pyramid is a square 12.0 cm on a side. The height of the pyramid is 18.5 cm. To find the volume we must realize that the base area B is $(12.0 \text{ cm})^2$. Thus,

$$V = \frac{1}{3}(12.0)^2(18.5)$$
$$= 888 \text{ cu cm.}$$

EXERCISES 4-3-A

In Exercises 1 and 2, find the volumes of the pyramids with polygons of base area B and height h.

1. $B = 4.00$ sq in., $h = 12.0$ in.
2. $B = 45.0$ sq m, $h = 22.0$ m

In Exercises 3 and 4, find the volumes of the pyramids with square bases of side s and height h.

3. $s = 9.00$ mm, $h = 17.0$ mm
4. $s = 36.0$ ft, $h = 7.25$ ft

4-3-5 In a regular pyramid the lateral faces are congruent isosceles triangles. The altitude, or height, of each of these triangles is the *slant height s* of the pyramid as shown in the figure at the right. Therefore, the area of each triangle is one-half of the product of the slant height and the base of the triangle. The sum of the bases of the triangles is the perimeter of the base of the pyramid. Thus, the *lateral area*, which is the sum of the areas of the lateral faces of the pyramid, is

$$L = \frac{1}{2}ps,$$

where p is the perimeter of the base and s is the slant height.

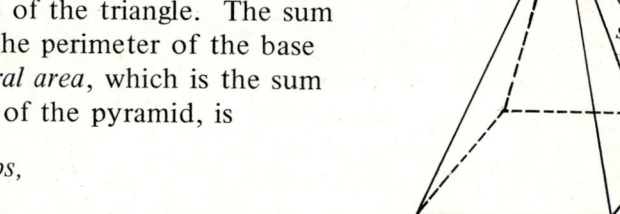

4-3-6 **EXAMPLE 4-3-B**

(a) The lateral area of a pyramid for which $p = 18.0$ cm and $s = 12.0$ cm is

$$L = \frac{1}{2}(18.0)(12.0)$$
$$= 108 \text{ sq cm.}$$

(b) To determine the lateral area of a pyramid with a square base of side 10.0 in. and height 12.0 in. we use the Pythagorean theorem to first find s. If we cut a section through the pyramid including h and s, we have the figure at the right. The base of the isosceles triangle shown equals the length of one of the sides of the square base. Thus, using one of the right triangles

$$s^2 = (5.00)^2 + (12.0)^2 = 169,$$

or

$s = 13.0$ in.

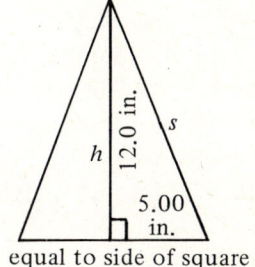
equal to side of square

Therefore, since the perimeter of the base is $4(10.0)$ in., we have

$$L = \frac{1}{2}(4)(10.0)(13.0)$$
$$= 260 \text{ sq in.}$$

EXERCISES 4-3-B

In Exercises 1 and 2, determine the lateral areas of the pyramids with the given values of p and s.

1. $p = 56.0$ m, $s = 21.0$ m
2. $p = 8.36$ ft, $s = 4.14$ ft

In Exercises 3 and 4, determine the lateral areas of the pyramids with square bases, with the given sides and given values of h.

3. side of square = 6.00 yd; $h = 4.00$ yd

4. side of square = 30.0 mm; $h = 8.00$ mm

4-3-7 A *conical surface* is generated by rotating a straight line about one of its points. A *cone* is a solid geometric figure bounded by the conical surface and a plane which cuts the surface. Thus, the base can be any curved geometric figure, and the conical surface can take on many configurations. Examples of different types of cones are shown at the right.

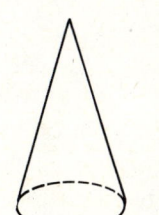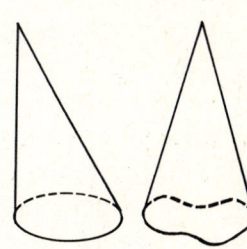

4-3-8 Although many types of cones may be generated, one type is of particular importance. It is the *right circular cone*, which is generated by rotating a right triangle about one of its legs. Thus, the *base* of the cone is a circle, and the *slant height* is the hypotenuse of the right triangle. The *height*, or *altitude*, of the cone is one leg of the right triangle, and the radius of the base is the other leg. We shall restrict our attention from this point to right circular cones, and whenever we use "cone" we shall mean "right circular cone."

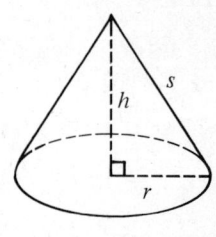

4-3-9 As we did with the prism and pyramid, if we compare the volumes of a cylinder and cone with the same base and height, we see that the volume of the cone is considerably less than that of the cylinder. In fact, the volume of the cone is exactly one-third of the volume of the cylinder, just as with the pyramid and prism. Thus, the volume of a right circular cone is

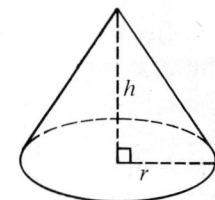

$$V = \frac{1}{3}\pi r^2 h,$$

where r is the radius of the base and h is the height of the cone.

4-3-10	**EXAMPLE 4-3-C**	**EXERCISES 4-3-C**
	The volume of a cone for which $r = 5.50$ in. and $h = 3.75$ in. is	Determine the volumes of the cones with the given values of r and h.
	$V = \frac{1}{3}\pi(5.50)^2(3.75)$	1. $r = 8.00$ cm, $h = 6.00$ cm
	$= \frac{1}{3}(3.14)(30.25)(3.75)$	2. $r = 1.60$ ft, $h = 2.40$ ft
	$= 119$ cu in.	3. $r = 30.0$ in., $h = 18.0$ in.
		4. $r = 0.875$ m, $h = 0.329$ m

4-3-11 It is reasonable that the lateral area of a cone follows the same basic formula as that of a pyramid. This is in fact the case. Recalling that the slant height of a cone is the hypotenuse of the right triangle with legs r and h, and that the perimeter of a circle is its circumference, we have

$$L = \frac{1}{2}ps = \frac{1}{2}(2\pi r)s,$$

or

$$L = \pi r s$$

as the formula for the lateral area of a cone.

4-3-12 **EXAMPLE 4-3-D**

A tent is in the shape of a cone with a radius of 3.20 m and height of 3.60 m. What is the surface area of the tent?

We must first find the slant height s. This is done by use of the Pythagorean theorem. As we noted above, the radius and height are the legs of a right triangle with the slant height as the hypotenuse. Thus,

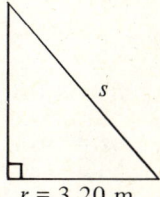

$$s^2 = r^2 + h^2 = (3.20)^2 + (3.60)^2$$
$$= 10.24 + 12.96 = 23.2,$$

or
$$s = 4.82 \text{ m}.$$

Thus, the lateral area is

$$L = \pi(3.20)(4.82) = (3.14)(3.20)(4.82)$$
$$= 48.4 \text{ sq m}.$$

EXERCISES 4-3-D

In Exercises 1 and 2, determine the lateral areas of the cones with the given values of r and s.

1. $r = 6.00$ ft, $s = 16.0$ ft
2. $r = 36.0$ cm, $s = 45.0$ cm

In Exercises 3 and 4, solve the given problems.

3. A sheet metal cover is in the shape of a cone with radius 17.0 cm and height 6.00 cm. What is the surface area of the cover?
4. A paper cup is in the shape of a cone with radius 1.80 in. and height 3.50 in. What is the surface area of the cup?

4-3-13 In this section we introduced pyramids and cones. We discussed the volume and lateral area of regular pyramids and right circular cones. The exercises which follow provide an opportunity to review the use of the formulas which were developed.

4-3-14 **EXERCISES 4-3-Section**

In Exercises 1-8, find the volumes of the pyramids with the given bases and heights.

1. Base is polygon with $B = 20.0$ sq m; $h = 12.0$ m.
2. Base is polygon with $B = 750$ sq in.; $h = 30.0$ in.
3. Base is polygon with $B = 3600$ sq ft; $h = 45.0$ ft.
4. Base is polygon with $B = 7850$ sq cm; $h = 38.4$ cm.
5. Base is square of side 25.0 mm; $h = 4.60$ mm.
6. Base is square of side 8.50 ft; $h = 4.85$ ft.
7. Base is square of side 2.35 yd; $h = 1.70$ yd.
8. Base is square of side 0.0123 km; $h = 0.0136$ km.

In Exercises 9-16, find the lateral areas of the pyramids with the given bases, slant heights, or heights.

9. Base is polygon with $p = 200$ mm; $s = 60.0$ mm.
10. Base is polygon with $p = 65.0$ ft; $s = 30.0$ ft.
11. Base is polygon with $p = 18.0$ yd; $s = 8.50$ yd.
12. Base is polygon with $p = 880$ cm; $s = 350$ cm.
13. Base is square of side 12.0 m; $h = 8.00$ m.
14. Base is square of side 30.0 in.; $h = 36.0$ in.
15. Base is equilateral triangle of side 23.2 ft; $s = 18.5$ ft.
16. Base is equilateral triangle of side 230 mm; $s = 165$ mm.

In Exercises 17-24, find the volumes of the cones with the given base radii or diameters, and heights.

17. $r = 10.0$ ft, $h = 14.0$ ft
18. $r = 25.0$ cm, $h = 40.0$ cm
19. $r = 2.30$ m, $h = 2.50$ m
20. $r = 6.80$ in., $h = 2.50$ in.
21. $d = 16.2$ yd, $h = 8.55$ yd
22. $d = 10.8$ m, $h = 14.3$ m
23. $d = 62.8$ cm, $h = 26.3$ cm
24. $d = 17.8$ ft, $h = 22.3$ ft

In Exercises 25-32, find the lateral areas of the cones with the given base radii or diameters, and slant heights or heights.

25. $r = 7.00$ in., $s = 9.00$ in.
26. $r = 80.0$ mm, $s = 120$ mm
27. $d = 2.80$ m, $s = 1.80$ m
28. $d = 36.0$ in., $s = 25.0$ in.
29. $r = 9.00$ ft, $h = 12.0$ ft
30. $r = 45.0$ cm, $h = 24.0$ cm
31. $d = 352$ mm, $h = 220$ mm
32. $d = 17.8$ ft, $h = 10.5$ ft

In Exercises 33-40, solve the given problems involving pyramids and cones.

33. When built, the Great Pyramid of Egypt had a square base that was approximately 250 yd on a side. Its height was about 160 yd. What was its volume?

34. A steel wedge is made in the shape of a pyramid with a square base, 3.50 cm on a side, and a height of 8.25 cm. Given that the density of steel is 7.80 grams per cubic centimeter, what is the weight of the wedge?

35. A tent has the shape of a pyramid on a square base. The side of the square is 2.80 m and the height of the pyramid is 3.15 m. What is the area of the canvas needed for the lateral sides and the floor?

36. A sheet metal container is in the shape of a pyramid (inverted) with a square base. The side of the square is 12.0 ft and the depth is 8.75 ft. At a cost of $0.75 per square foot, what is the cost of the material of the container (no base)?

37. A conical cistern 10.0 ft high has a radius at the top of 7.50 ft. Given that the density of water is 62.4 pounds per cubic foot, what weight of water can the cistern hold?

38. A pile of sand is in the shape of a cone. The diameter of the base is 12.0 m and its height is 2.50 m. What is the volume of sand in the pile?

39. A conical funnel is 8.50 cm deep and 10.0 cm in diameter. What is the surface area of the funnel? (Neglect the opening of the funnel.)

40. An umbrella opens into the shape of a cone 3.00 ft in diameter and 0.500 ft deep. What is the surface area of the umbrella?

4-4 THE SPHERE

4-4-1 The final solid figure we consider is the *sphere*. Although there are numerous applications of the sphere in areas such as architecture, engineering, and astronomy, one of the most practical reasons for studying the sphere is that the earth is essentially spherical.

4-4-2 A sphere is generated if a circle is rotated about a diameter. Every point on the sphere is equidistant from the *center*. The *radius* of a sphere is a line segment joining the center and a point on the sphere. The *diameter* of a sphere is a line segment through the center and having its end points on the sphere. The intersection of a plane and sphere is a *circle*. If the plane of intersection contains the center, the circle of intersection is a *great circle*. Other circles in planes not containing the center are *small circles*. These are illustrated in the figure at the right.

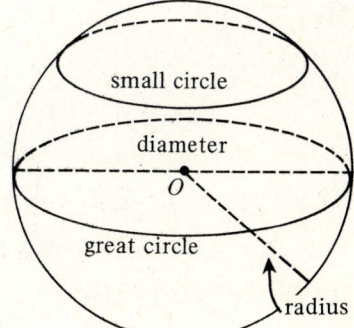

88   Solid Geometric Figures

4-4-3   Due to the close relationship between the circle and sphere, the volume of a sphere is expressed in terms of π. Although we cannot prove it here, the volume of a sphere is

$$V = \frac{4}{3}\pi r^3,$$

where r is the radius of the sphere.

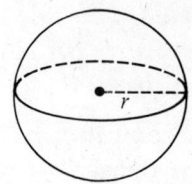

4-4-4   **EXAMPLE 4-4-A**

(a) The volume of a sphere of radius 8.00 cm is

$$V = \frac{4}{3}\pi(8.00)^3 = \frac{4(3.14)(512)}{3}$$

$$= 2140 \text{ cu cm}.$$

(b) The volume of a sphere of diameter 6.80 in. is

$$V = \frac{4}{3}\pi(3.40)^3 = \frac{4(3.14)(39.3)}{3}$$

$$= 165 \text{ cu in}.$$

EXERCISES 4-4-A

Find the volumes of the spheres with the given radii and diameters.

1. $r = 6.00$ m
2. $r = 15.0$ ft
3. $d = 4.20$ in.
4. $d = 86.0$ cm

4-4-5   The surface area of a sphere is also measured in terms of π. The formula for the surface area of a sphere is

$$A = 4\pi r^2,$$

where r is the radius of the sphere.

4-4-6   **EXAMPLE 4-4-B**

(a) The surface area of a sphere of radius 8.00 mm is

$$A = 4\pi(8.00)^2 = 4(3.14)(64.0)$$

$$= 804 \text{ sq mm}.$$

(b) The surface area of a sphere of diameter 36.0 ft is

$$A = 4\pi(18.0)^2 = 4(3.14)(324)$$

$$= 4070 \text{ sq ft}.$$

EXERCISES 4-4-B

Find the surface areas of the spheres with the given radii and diameters.

1. $r = 9.00$ m   2. $r = 18.0$ ft
3. $d = 84.0$ ft   4. $d = 0.860$ cm

4-4-7   The following example illustrates the use of a sphere in combination with a cylinder in an applied situation.

4-4-8 **EXAMPLE 4-4-C**

A grain storage container is in the shape of a cylinder surmounted by a hemisphere (half a sphere). Given that the radius of the cylinder is 40.0 ft and its height is 120 ft, find (a) the volume of the container and (b) the external surface area.

(a) The volume of the container is the volume of the cylinder plus the volume of the hemisphere. By the construction we see that the radius of the hemisphere is the same as the radius of the cylinder. Thus,

$$V = \pi r^2 h + \frac{1}{2}\left(\frac{4}{3}\pi r^3\right)$$

$$= \pi r^2 h + \frac{2}{3}\pi r^3$$

$$= (3.14)(40.0)^2(120) + \frac{2(3.14)(40.0)^3}{3}$$

$$= 603{,}000 + 134{,}000 = 737{,}000 \text{ cu ft.}$$

(b) The external surface area of the container is the lateral area of the cylinder plus the surface area of the hemisphere. Thus,

$$A = 2\pi rh + \frac{1}{2}(4\pi r^2) = 2\pi rh + 2\pi r^2$$

$$= 2(3.14)(40.0)(120) + 2(3.14)(40.0)^2$$

$$= 30{,}100 + 10{,}000 = 40{,}100 \text{ sq ft.}$$

EXERCISES 4-4-C

A storage container has the same shape (cylinder and hemisphere) as that in Example 4-4-C. Given that the radius of the cylinder is 30.0 ft and the height of the cylinder is 80.0 ft, find

1. the volume, and
2. the external surface area of the container.

4-4-9 In this section we introduced the sphere along with formulas for its volume and surface area. The following exercises provide an opportunity to review the use of these formulas.

4-4-10 EXERCISES 4-4-Section

In Exercises 1-8, find the volumes of the spheres with the given radii or diameters.

1. $r = 3.00$ ft
2. $r = 20.0$ cm
3. $r = 1.40$ m
4. $r = 58.0$ in.
5. $d = 14.0$ yd
6. $d = 220$ mm
7. $d = 4.68$ cm
8. $d = 36.2$ ft

In Exercises 9-16, find the surface areas of the spheres with the given radii or diameters.

9. $r = 30.0$ in.
10. $r = 60.0$ m
11. $r = 3.40$ cm
12. $r = 46.0$ in.
13. $d = 0.500$ mi
14. $d = 0.300$ km
15. $d = 346$ mm
16. $d = 62.4$ in.

In Exercises 17-22, solve the given problems involving spheres.

17. A tennis ball is 3.34 cm in radius. What is the volume of a tennis ball?

18. What is the weight of a gold sphere 6.00 in. in diameter? The density of gold is 0.697 pounds per cubic inch.

19. The radius of the earth is 3960 mi. What is the surface area of the earth?

20. A lampshade is in the shape of a hemisphere of diameter 36.0 cm. What is the surface area of the shade?

21. A gasoline storage tank is in the shape of a cylinder with hemispheres at each end. The cylindrical part is 6.50 m long and 4.40 m in diameter. What is the capacity, in liters, of the tank? (1 cu m = 1000 liters.)

22. A hollow metal sphere has an outer radius of 1.50 ft and an inner radius of 1.25 ft. What is the volume of the metal in the sphere?

23. What is the radius of a 16.0-lb shot? The density of iron used in the shot is 450 pounds per cubic foot.

24. The surface area of a basketball is 1820 sq cm. What is the radius of the basketball?

4-5 EXERCISES FOR UNIT FOUR

In Exercises 1-20, find the volume of the indicated figure for the given values.

1. Prism: base area 40.0 sq ft, height 15.0 ft
2. Prism: base area 800 sq cm, height 50.0 cm
3. Rectangular solid: length 2.00 m, width 1.50 m, height 1.20 m

4. Rectangular solid: length 25.0 in., width 10.0 in., height 15.0 in.
5. Cube: edge 3.5 yd
6. Cube: edge 220 mm
7. Prism: base is right triangle with legs 26.0 cm and 34.0 cm; height 14.0 cm
8. Prism: base is right triangle with legs 6.20 ft and 3.80 ft; height 8.50 ft
9. Cylinder: base radius 36.0 in., altitude 24.0 in.
10. Cylinder: base radius 2.80 m, altitude 1.50 m
11. Cylinder: base diameter 120 mm, height 58.0 mm
12. Cylinder: base diameter 16.2 yd, height 12.4 yd
13. Pyramid: base area 330 sq ft, height 25.0 ft
14. Pyramid: base area 3850 sq cm, height 125 cm
15. Pyramid: base is square with side 13.2 m, altitude 7.50 m
16. Pyramid: base is square with side 6.44 in., altitude 2.25 in.
17. Cone: base radius 18.2 ft, height 11.5 ft
18. Cone: base diameter 76.2 mm, height 22.1 mm
19. Sphere: radius 18.0 cm
20. Sphere: diameter 52.0 ft

In Exercises 21-40, find the lateral area or total area of the indicated figure for the given values.

21. Lateral area of prism: base perimeter 60.0 ft, height 15.0 ft
22. Lateral area of prism: base perimeter 130 cm, height 50.0 cm
23. Total area of rectangular solid of Exercise 3
24. Total area of rectangular solid of Exercise 4
25. Total area of cube of Exercise 5
26. Total area of cube of Exercise 6
27. Total area of prism of Exercise 7
28. Total area of prism of Exercise 8
29. Lateral area of cylinder of Exercise 9
30. Lateral area of cylinder of Exercise 10

31. Total area of cylinder of Exercise 11
32. Total area of cylinder of Exercise 12
33. Lateral area of pyramid: base perimeter 85.0 ft, slant height 32.0 ft
34. Lateral area of pyramid: base perimeter 240 cm, slant height 140 cm
35. Lateral area of pyramid of Exercise 15
36. Lateral area of pyramid of Exercise 16
37. Lateral area of cone of Exercise 17
38. Lateral area of cone of Exercise 18
39. Total area of sphere of Exercise 19
40. Total area of sphere of Exercise 20

In Exercises 41-56, solve the given problems involving the solid figures of this unit.

41. A rectangular brass bar is 6.00 in. long, 2.50 in. wide, and 2.00 in. high. The density of brass is 0.295 pounds per cubic inch. What is the weight of the bar?
42. A boxcar is 11.2 m long, 2.50 m wide, and can be filled to a depth of 1.70 m. What volume of material can the boxcar hold?
43. A retaining wall 1.50 m high is built around a rectangular piece of land 36.0 m long and 32.0 m wide. What is the area of the wall?
44. A cubical box with an outside edge of 8.00 in. is made of wood 0.75 in. thick. What is the actual volume of the inside of the box?
45. The tank on an oil tank car is a cylinder 38.5 ft long and 7.10 ft in diameter. What is its capacity in gallons? One cubic foot contains 7.48 gallons.
46. An iron water pipe has an inside diameter of 2.22 cm and an outside diameter of 2.86 cm. Find the weight of 100 cm of pipe if the density of iron is 7.80 grams per cubic centimeter.
47. A furnace is built in the shape of a cylinder 1.40 m high and 0.950 m in diameter. How many square meters of insulation are required to cover the sides and top of the furnace?
48. Compare the volumes and total surface areas of two cylindrical containers. One is 4.00 in. in radius and 8.00 in. high, and the other is 8.00 in. in radius and 2.00 in. high.

49. A certain type of rivet is shaped as shown in the figure at the right (a conical part on a cylindrical part). Find the volume of the rivet shown.

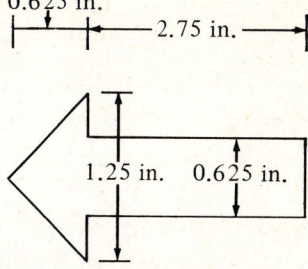

50. A marble monument is in the shape of a pyramid with a square base 160 cm on a side. The height of the monument is 140 cm. What is the weight, in kilograms, of the monument? The density of marble is 0.0270 kilograms per cubic centimeter.

51. Filter paper lines the inside of a conical funnel 16.0 cm in radius and 12.5 cm deep. What area of filter paper is needed?

52. What is the volume of a conical rod with a base diameter of 0.750 in. and 48.0 in. long?

53. The radius of a baseball is 1.43 in. What is its volume?

54. The diameter of the moon is 2160 mi. What is the surface area of the moon?

55. By submerging a prism with an irregular base in water, its volume is found to be 1200 cu cm. If the height of the prism is 26.5 cm, what is its base area?

56. A silver sphere weighs 1250 gm. The density of silver is 10.5 grams per cubic centimeter. What is the radius of the sphere?

A Note Regarding Units of Measurement

Presently there are two major systems of measurement, the English system and the metric system. However, the major countries which use the English system have started the conversion to the metric system. There is a great deal of evidence of this transition in American industry. The measurements primarily affected by this change are those of length, area, volume, and weight. Therefore, since we are in this state of transition, about one-half of the applied problems in this text dealing with these measurements are in the English system and the other half are in the metric system.

For the information and convenience of the reader, we present here a very brief summary of the basic units of the English system and the metric system, along with a few basic conversions between the systems. The indicated abbreviations are used in the text.

	English	Metric
Length:	1 yard (yd) = 3 feet (ft) 12 inches (in) = 1 ft 1 mile (mi) = 5280 ft	100 centimeters (cm) = 1 meter (m) 1000 millimeters (mm) = 1 m 1000 m = 1 kilometer (km)
Volume:	4 quarts (qt) = 1 gallon (gal)	1000 milliliters (ml) = 1 liter (l) 1000 l = 1 kiloliter (kl)
Weight:	16 ounces (oz) = 1 pound (lb) 1 ton = 2000 lb	1000 milligrams (mg) = 1 gram (g) 1000 g = 1 kilogram (kg)

Basic Conversions

1 in = 2.54 cm   1 l = 1.06 qt
1 m = 39.4 in   1 kg = 2.2 lb

It can be seen that 1 cm is somewhat less than one-half an inch, and that a meter is a little longer than a yard. A liter is almost the same as a quart, and a kilogram is a little more than two pounds. These conversion factors should give the reader a sense of the magnitude of some of the basic metric units.

When using the metric system, temperature is normally measured in degrees centigrade. It is probable that °C will become universally used with the metric system. Briefly, 0°C is the temperature at which ice melts, and 100°C is the temperature at which water boils. These are equivalent to 32°F and 212°F (Fahrenheit) respectively.

This information on units is intended primarily to give to the reader some meaning of the metric units which appear in some of the exercises relevant to applied situations. No changing of units is needed with the exercises.

SOLUTIONS FOR ALL EXERCISES OF SHORT EXERCISE SETS
and
ANSWERS FOR ODD-NUMBERED EXERCISES OF SECTION AND UNIT EXERCISES
UNIT ONE

EXERCISES 1-1-A

1. $\angle POQ$, $\angle QOP$ 2. point O 3. OP, OQ

EXERCISES 1-1-B

1. $360°$ 2. $180°$ (straight angle)

EXERCISES 1-1-C

1. $PQ \perp OR$ 2. $\angle POR$, $\angle QOR$

EXERCISES 1-1-D

1. $25°$ 2. $130°$

EXERCISES 1-1-E

1. $0.3° = (0.3)(60') = 18'$; $41.3° = 41°18'$ 2. $0.75° = (0.75)(60') = 45'$; $12.75° = 12°45'$
3. $42' = \left(\frac{42}{60}\right)° = 0.7°$; $86°42' = 86.7°$ 4. $39' = \left(\frac{39}{60}\right)° = 0.65°$; $7°39' = 7.65°$

EXERCISES 1-1-F

1. acute ($83°$ is between $0°$ and $90°$)
   obtuse ($102°$ is between $90°$ and $180°$)
   obtuse ($90°30'$ is between $90°$ and $180°$)

2. obtuse ($160°$ is between $90°$ and $180°$)
   acute ($89°59'$ is between $0°$ and $90°$)
   acute ($9°$ is between $0°$ and $90°$)

EXERCISES 1-1-Section

1. $\angle BCD$, $\angle DCB$ 3. $\angle ABC$ 5. DB, AB
7. $\angle CBD$ 9. DA and AC 11. $\angle BAD$
13. $30°$, acute   15. $110°$, obtuse   17. $115°$, obtuse   19. $70°$, acute
21. $56°24'$ 23. $136°27'$ 25. $156.25°$ 27. $67.1°$

29.

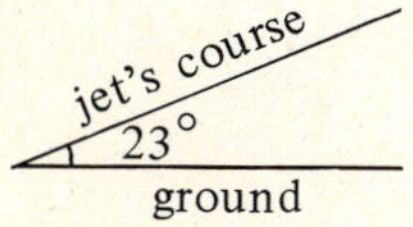

jet's course
$23°$
ground

95

96 Solutions

EXERCISES 1-2-A

1. ∠POQ and ∠QOR (vertex at O, common side OQ)

2. (one possibility)

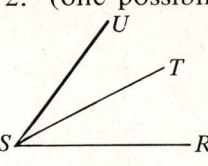

(vertex at S, common side ST)

EXERCISES 1-2-B

1. ∠POQ and ∠ROS; ∠POS and ∠QOR

2. (one possibility)

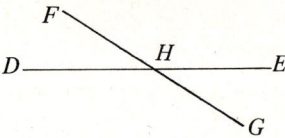

∠FHD and ∠GHE; ∠EHF and ∠DHG

EXERCISES 1-2-C

1. ∠RST and ∠XYZ are supplementary since 70° + 110° = 180°

2. (one possibility)

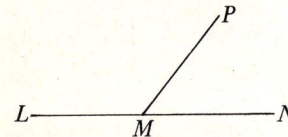

∠NMP and ∠PML are supplementary since ∠NMP + ∠PML = ∠NML = 180°

EXERCISES 1-2-D

1. ∠RST and ∠UVW are complementary since 70° + 20° = 90°

2. (one possibility)

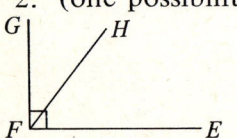

∠EFH and ∠HFG are complementary since ∠EFH + ∠HFG = ∠EFG = 90°

EXERCISES 1-2-E

1. 121°; 180° − 59° = 121°
2. 31°; 90° − 59° = 31°
3. ∠EBC = 180° − 76° = 104°
4. ∠CBD = 90° − 76° = 14°

EXERCISES 1-2-F

1. ∠2 and ∠6, or ∠3 and ∠7, or ∠4 and ∠8
2. ∠3 and ∠6

EXERCISES 1-2-Section

1. $\angle AED$ and $\angle BEC$; $\angle AEB$ and $\angle CED$
3. $\angle AEB$ and $\angle BEC$; $\angle BEC$ and $\angle CED$; $\angle AEC$ and $\angle CED$; $\angle AEB$ and $\angle BED$ (any two pairs)
5. $\angle CBE$ and $\angle EBA$ 7. $\angle CBD$ 9. 25°   11. 115°
13. 40°   15. 62°   17. $\angle 1$ and $\angle 5$; $\angle 3$ and $\angle 4$
19. $\angle 1$ and $\angle 3$, or $\angle 4$ and $\angle 5$ 21. 50°   23. 130°   25. 58°   27. 148°

EXERCISES 1-3-A

1. $\triangle LMN$ ($LN = MN$)   2. $\triangle RST$ (no equal sides)
3. $\triangle CDE$ ($CD = DE = EC$)   4. 60° (equilateral triangle)
5. $\angle L$ and $\angle M$ 6. 40° (base angles are equal)

EXERCISES 1-3-B

1. LN (opposite $\angle M$)   2. LM and MN

EXERCISES 1-3-C

1. $\angle B + \angle C = 42° + 75° = 117°$; $\angle A = 180° - 117° = 63°$
2. $\angle B + \angle C = 72° + 69° = 141°$; $\angle A = 180° - 141° = 39°$
3. $\angle B + \angle C = 63° + 90° = 153°$; $\angle A = 180° - 153° = 27°$
4. $\angle B + \angle C = 17° + 90° = 107°$; $\angle A = 180° - 107° = 73°$

EXERCISES 1-3-D

1. $\angle ACB = 115°$ (vertical angles); $\angle B + \angle ACB = 20° + 115° = 135°$; $\angle A = 180° - 135° = 45°$
2. $\angle ABC = 180° - 143° = 37°$; $\angle ABC + \angle C = 37° + 90° = 127°$; $\angle A = 180° - 127° = 53°$

EXERCISES 1-3-Section

1. $\triangle DBC$ 3. $\triangle ADC$ 5. 5   7. 120°   9. 53°   11. 38°
13. 40°   15. 40°   17. 33°   19. 124°   21. 118°   23. No; no; yes

EXERCISES 1-4-A

1. $BC = 3$ cm (opposite sides equal)
2. $CD = 5$ cm (opposite sides equal)
3. $\angle C = 65°$ (opposite angles equal)
4. $\angle D = 115°$ (opposite angles equal)
5. $FG = 4$ in. (all sides equal)
6. $EH = 4$ in. (all sides equal)
7. $\angle G = 72°$ (opposite angles equal)
8. $\angle H = 108°$ (opposite angles equal)

EXERCISES 1-4-B

1. $BC = 2$ in. (opposite sides equal)
2. $CD = 5$ in. (opposite sides equal)
3. $FG = 3$ cm (all sides equal)
4. $\angle H + \angle E = 90° + 90° = 180°$

EXERCISES 1-4-C

1. EF and HG (the parallel sides)
2. $\angle G$ and $\angle H$ (of HG); $\angle E$ and $\angle F$ (of EF)

EXERCISES 1-4-D

1. $\angle B + \angle C + \angle D = 72° + 90° + 110° = 272°$; $\angle A = 360° - 272° = 88°$
2. $\angle B + \angle C + \angle D = 68° + 80° + 122° = 270°$; $\angle A = 360° - 270° = 90°$
3. $\angle C = 74°$ ($\angle C = \angle A$); $\angle A + \angle C = 74° + 74° = 148°$; $2(\angle B) = 360° - 148° = 212°$; $\angle B = 106°$; $\angle D = 106°$ ($\angle D = \angle B$)
4. $\angle C = 134°$ ($\angle C = \angle A$); $\angle A + \angle C = 134° + 134° = 268°$; $2(\angle B) = 360° - 268° = 92°$; $\angle B = 46°$; $\angle D = 46°$ ($\angle D = \angle B$)

EXERCISES 1-4-Section

1. (one possibility)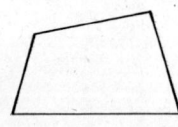
3.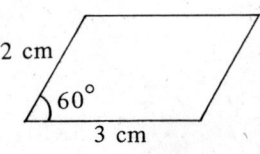
5.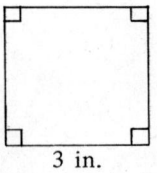
7.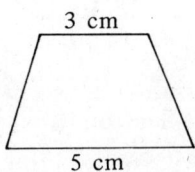
9. 70°

11. 110°
13. $\angle B = 100°$, $\angle C = 80°$, $\angle D = 100°$
15. $\angle B = 65°$, $\angle C = 115°$, $\angle D = 65°$
17. $\angle B = 85°$, $\angle C = 95°$, $\angle D = 95°$
19. $\angle B = 62°$, $\angle C = 118°$, $\angle D = 118°$
21. $\angle A = 60°$, $\angle B = 120°$, $\angle C = 60°$, $\angle D = 120°$
23. $\angle A = 70°$, $\angle B = 110°$, $\angle C = 70°$, $\angle D = 110°$
25. $90° + 90° + 90° + 90° = 360°$
27. perpendicular

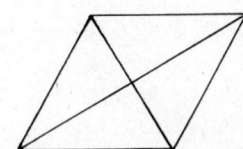

Solutions 99

EXERCISES FOR UNIT ONE (Section 1-5)

1. ∠ABD, ∠CBD   3. ∠CBE   5. ∠EBD   7. ∠CBD and ∠DBE
9. ∠2 and ∠5   11. ∠3 and ∠4   13. △ABD, △BCD   15. ABDE
17. 43°57′   19. 105.9°   21. 65°   23. 32°   25. 52°
27. 132°   29. 36°   31. 31°   33. 96°   35. 50°

37. since corresponding angles are equal, the lines are parallel.

39. ∠C = ∠ABE = 60° (corresponding angles); ∠D = ∠BEA = 60° (corresponding angles)
∠A = 60°; all angles have measure of 60°.

41 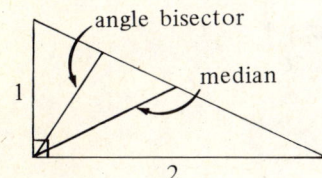   43. 34°

UNIT TWO

EXERCISES 2-1-A

1. $p = 2$ ft $+ 3$ ft $+ 4$ ft $= 9$ ft   2. $p = 18$ m $+ 24$ m $+ 30$ m $= 72$ m
3. $p = 12$ mm $+ 29$ mm $+ 14$ mm $+ 17$ mm $= 72$ mm
4. $p = 9$ in. $+ 53$ in. $+ 12$ in. $+ 38$ in. $+ 19$ in. $= 131$ in.

EXERCISES 2-1-B

1. $p = 35$ in. $+ 41$ in. $+ 52$ in. $= 128$ in.   2. $p = 521$ mm $+ 376$ mm $+ 296$ mm $= 1193$ mm
3. $p = 79$ yd $+ 63$ yd $+ 46$ yd $+ 91$ yd $= 279$ yd
4. $p = 317$ m $+ 142$ m $+ 256$ m $+ 407$ m $= 1122$ m

EXERCISES 2-1-C

1. $p = 3(14$ m$) = 42$ m   2. $p = 3(520$ yd$) = 1560$ yd
3. $p = 2(26$ cm$) + 45$ cm $= 52$ cm $+ 45$ cm $= 97$ cm
4. $p = 2(37$ in.$) + 19$ in. $= 74$ in. $+ 19$ in. $= 93$ in.

100 Solutions

EXERCISES 2-1-D

1. $p = 4(17 \text{ ft}) = 68 \text{ ft}$
2. $p = 4(46 \text{ mm}) = 184 \text{ mm}$
3. $p = 2(29 \text{ cm}) + 2(17 \text{ cm}) = 58 \text{ cm} + 34 \text{ cm} = 92 \text{ cm}$
4. $p = 2(15 \text{ in.}) + 2(9 \text{ in.}) = 30 \text{ in.} + 18 \text{ in.} = 48 \text{ in.}$

EXERCISES 2-1-E

1.

   6 ft, 5 ft, 14 ft − 6 ft = 8 ft, 4 ft, 14 ft, 4 ft + 5 ft = 9 ft

   $p = 14 \text{ ft} + 4 \text{ ft} + 8 \text{ ft} + 5 \text{ ft} + 6 \text{ ft} + 9 \text{ ft}$
   $= 46 \text{ ft}$

2.

   140 cm, 52 cm, 140 cm, 140 cm, 140 cm

   $p = 140 \text{ cm} + 140 \text{ cm} + 140 \text{ cm} + 140 \text{ cm} + 52 \text{ cm}$
   $= 612 \text{ cm}$
   (length of dashed line not included)

EXERCISES 2-1-F

1. $p = 2(120 \text{ m}) + 2(105 \text{ m}) = 240 \text{ m} + 210 \text{ m} = 450 \text{ m}$
   450 m of fencing

2. perimeter of one table: $p = 4(33 \text{ in.}) = 132 \text{ in.}$
   perimeter of two tables: $2p = 2(132 \text{ in.}) = 264 \text{ in.} = \dfrac{264 \text{ in.}}{12 \text{ in./ft}} = 22 \text{ ft}$
   Cost at 7 cents/ft: $C = (22 \text{ ft})(7 \text{ cents/ft}) = 154 \text{ cents} = \$1.54$

EXERCISES 2-1-Section

1. 12 ft
3. 123 mm
5. 896 m
7. 84.8 in.
9. 193.8 mm
11. 57.1 in.
13. 2.6 m
15. 0.60 mi
17. 168.0 cm
19. 230.6 ft
21. 23.4 ft
23. 3630 cm
25. 21.22 yd
27. 150.9 mm
29. $18.72
31. 88.8 cm

EXERCISES 2-2-A

1. $A = (16 \text{ in.})(10 \text{ in.}) = 160 \text{ sq in.}$
2. $A = (20 \text{ m})(14 \text{ m}) = 280 \text{ sq m}$
3. $A = (120 \text{ cm})^2 = 14{,}400 \text{ sq cm}$
4. $A = (6.5 \text{ ft})^2 = 42.25 \text{ sq ft}$

EXERCISES 2-2-B

1. $A = (3.5 \text{ ft})(4.0 \text{ ft}) = 14$ sq ft
2. $A = (28 \text{ mm})(20 \text{ mm}) = 560$ sq mm
3. $A = (65 \text{ cm})(26 \text{ cm}) = 1690$ sq cm
4. $A = (3.2 \text{ yd})(3.8 \text{ yd}) = 12.16$ sq yd

EXERCISES 2-2-C

1. $A = \frac{1}{2}(3.9 \text{ m})(4.2 \text{ m}) = 8.19$ sq m
2. $A = \frac{1}{2}(29 \text{ ft})(18 \text{ ft}) = 261$ sq ft
3. $A = \frac{1}{2}(17.4 \text{ in.})(11.2 \text{ in.}) = 97.44$ sq in.
4. $A = \frac{1}{2}(310 \text{ mm})(385 \text{ mm}) = 59{,}675$ sq mm

EXERCISES 2-2-D

1. $A = \frac{1}{2}(6 \text{ in.})(15 \text{ in.} + 9 \text{ in.}) = \frac{1}{2}(6 \text{ in.})(24 \text{ in.}) = 72$ sq in.
2. $A = \frac{1}{2}(1.6 \text{ cm})(3.4 \text{ cm} + 2.7 \text{ cm}) = \frac{1}{2}(1.6 \text{ cm})(6.1 \text{ cm}) = 4.88$ sq cm
3. $A = \frac{1}{2}(24 \text{ m})(76 \text{ m} + 62 \text{ m}) = \frac{1}{2}(24 \text{ m})(138 \text{ m}) = 1656$ sq m
4. $A = \frac{1}{2}(1.2 \text{ ft})(4.8 \text{ ft} + 3.4 \text{ ft}) = \frac{1}{2}(1.2 \text{ ft})(8.2 \text{ ft}) = 4.92$ sq ft

EXERCISES 2-2-E

1. $A = (14.0 \text{ ft})(8.0 \text{ ft}) - (3.5 \text{ ft})^2 = 112.0$ sq ft $- 12.25$ sq ft $= 99.75$ sq ft
2. $A = (17.0 \text{ cm})^2 + \frac{1}{2}(16.0 \text{ cm})(17.0 \text{ cm} + 8.5 \text{ cm}) = 289$ sq cm $+ 204$ sq cm $= 493$ sq cm;
   $C = (\$0.85/\text{sq cm})(493 \text{ sq cm}) = \$419.05$

EXERCISES 2-2-Section

1. 2700 sq cm
3. 57.76 sq in.
5. 2448 sq mm
7. 105.6 sq in.
9. 0.24 sq m
11. 237.6 sq ft
13. 0.000258 sq km
15. 18,870 sq in.
17. (a) 185.5 sq in. (b) 58.6 in.
19. (a) 14.193 sq m (b) 17.65 m
21. 1024.1 sq cm
23. 488.1 sq yd
25. 1044 sq in.
27. $115.25

EXERCISES 2-3-A

1. $c^2 = 5^2 + 12^2 = 25 + 144 = 169;\ c = \sqrt{169} = 13$
2. $c^2 = 16^2 + 30^2 = 256 + 900 = 1156;\ c = \sqrt{1156} = 34$
3. $c^2 = 6^2 + 9^2 = 36 + 81 = 117;\ c = \sqrt{117} = 10.8$
4. $c^2 = 12^2 + 17^2 = 144 + 289 = 433;\ c = \sqrt{433} = 20.8$

EXERCISES 2-3-B

1. $a^2 = 15^2 - 9^2 = 225 - 81 = 144$; $a = \sqrt{144} = 12$
2. $b^2 = 26^2 - 16^2 = 676 - 256 = 420$; $b = \sqrt{420} = 20.5$
3. $b^2 = 2.4^2 - 1.8^2 = 5.76 - 3.24 = 2.52$; $b = \sqrt{2.52} = 1.59$
4. $a^2 = 350^2 - 230^2 = 122{,}500 - 52{,}900 = 69{,}600$; $a = \sqrt{69{,}600} = 264$

EXERCISES 2-3-C

1.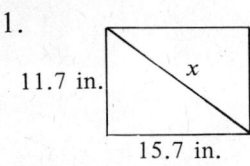

$x^2 = 11.7^2 + 15.7^2$
$= 136.89 + 246.49$
$= 383.38$
$x = \sqrt{383.38} = 19.6$ in.

2.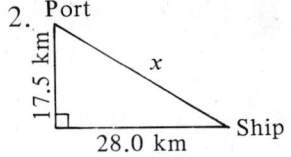

$x^2 = 17.5^2 + 28.0^2$
$= 306.25 + 784.00$
$= 1090.25$
$x = \sqrt{1090.25} = 33.0$ km

EXERCISES 2-3-D

1.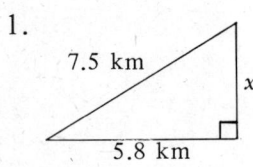

$x^2 = 7.5^2 - 5.8^2$
$= 56.25 - 33.64$
$= 22.61$
$x = \sqrt{22.61} = 4.75$ km

2.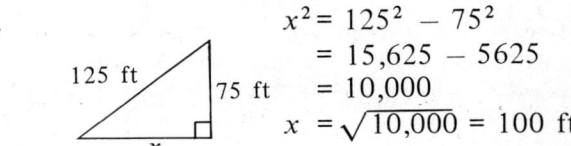

$x^2 = 125^2 - 75^2$
$= 15{,}625 - 5625$
$= 10{,}000$
$x = \sqrt{10{,}000} = 100$ ft

EXERCISES 2-3-Section

1. 5   3. 21   5. 15   7. 11.2   9. 44.5   11. 56
13. 21.1 ft   15. 510 m   17. 521 ft   19. 4.73 m

EXERCISES 2-4-A

1. Corresponding angles: $\angle A$ and $\angle D$; $\angle B$ and $\angle E$; $\angle C$ and $\angle F$
   Corresponding sides: AB and DE; BC and EF; AC and DF

2. Corresponding angles: $\angle R$ and $\angle V$; $\angle S$ and $\angle U$; $\angle T$ and $\angle W$
   Corresponding sides: RS and VU; ST and UW; RT and VW

3. Corresponding angles: $\angle D$ and $\angle G$; $\angle E$ and $\angle J$; $\angle F$ and $\angle H$
   Corresponding sides: DE and GJ; EF and JH; DF and GH

4. Corresponding angles: $\angle P$ and $\angle X$; $\angle Q$ and $\angle Z$; $\angle R$ and $\angle Y$
   Corresponding sides: PQ and XZ; QR and ZY; PR and XY

EXERCISES 2-4-B

1. $\dfrac{AB}{DE} = \dfrac{AC}{DF}$; $\dfrac{4}{8} = \dfrac{2}{DF}$; $DF = 4$

2. $\dfrac{RS}{UV} = \dfrac{ST}{UW}$; $\dfrac{9}{6} = \dfrac{12}{UW}$; $UW = 8$

3. $\dfrac{FE}{JH} = \dfrac{FD}{GH}$; $\dfrac{24}{18} = \dfrac{FD}{15}$; $FD = 20$

4. $\dfrac{PR}{XY} = \dfrac{RQ}{YZ}$; $\dfrac{4}{7} = \dfrac{RQ}{14}$; $RQ = 8$

EXERCISES 2-4-C

1. $\dfrac{PR}{SU} = \dfrac{PQ}{ST} = \dfrac{QR}{TU}$; $\dfrac{PR}{12} = \dfrac{14}{8} = \dfrac{QR}{16}$; $\dfrac{14}{8} = \dfrac{QR}{16}$; $QR = 28$

2. $\dfrac{PR}{SU} = \dfrac{PQ}{ST} = \dfrac{QR}{TU}$; $\dfrac{PR}{12} = \dfrac{14}{8} = \dfrac{QR}{16}$; $\dfrac{PR}{12} = \dfrac{14}{8}$; $PR = 21$

EXERCISES 2-4-D

1. $\dfrac{AD}{AB} = \dfrac{AE}{AC}$; $\dfrac{14}{20} = \dfrac{AE}{30}$; $AE = 21$ in.

2.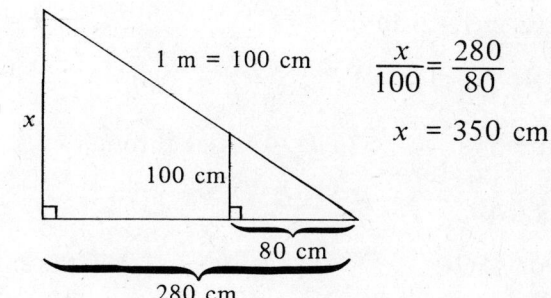
$\dfrac{x}{100} = \dfrac{280}{80}$
$x = 350$ cm

EXERCISES 2-4-E

1. On drawing $BC \equiv 1.25$ in.; let x represent actual distance BC;

$\dfrac{x}{1.25 \text{ in.}} = \dfrac{4 \text{ ft}}{1 \text{ in.}}$; $x = 5$ ft

2. On drawing $BC = 1.25$ in. and $AB = 0.75$ in.; let x represent actual distance BC;

$\dfrac{3 \text{ ft}}{0.75 \text{ in.}} = \dfrac{x}{1.25 \text{ in.}}$; $x = 5$ ft

EXERCISES 2-4-Section

1. $\angle E$; side EF 3. $\angle U$; side ST 5. 10   7. 12   9. 21
11. 20   13. 16   15. 3   17. 27 ft   19. 1600 cm
21. 3.50 ft   23. 38 m

EXERCISES FOR UNIT TWO (Section 2-5)

1. 40.2 in.   3. 27.2 m   5. 1.29 yd   7. 278 cm   9. 1798 ft
11. 153 mm   13. 224 sq cm   15. 3.082 sq yd   17. 4.797 sq cm
19. 3.055 sq in.   21. 0.00148 sq km   23. 1196 sq mm   25. 62.0 cm
27. 1232 ft   29. 156 sq cm   31. 77,000 sq ft   33. 41   35. 42
37. 7.36   39. 21.1   41. $\frac{39}{7}$ 43. $\frac{35}{3}$ 45. $8820   47. 153.6 sq m
49. 42.4 ft   51. 21.0 cm   53. $\frac{16}{3}$ ft   55. 1.50 m

EXERCISES 3-1-A

1. OS (or SO), OT (or TO)   2. ST (or TS)
3. $ST = 2(2 \text{ in.}) = 4$ in.   4. $r = \frac{1}{2}d = \frac{1}{2}(5 \text{ m}) = \frac{5}{2}$ m

EXERCISES 3-1-B

1. RS and MS 2. PQ (passes through)   3. NL (touches only at S)   4. RS

EXERCISES 3-1-C

1. $\angle POQ$ or $\angle SOR$ 2. $\overset{\frown}{RS}$ ($\overset{\frown}{STR}$ is the major arc)
3. $\overset{\frown}{SR} = 80°$ (measure equals that of central angle)   4. $\overset{\frown}{STR} = 360° - 80° = 280°$

EXERCISES 3-1-D

1. $\angle CDE$; $\overset{\frown}{CE}$ 2. $\angle CDE = \frac{1}{2}\overset{\frown}{CE} = \frac{1}{2}(84)° = 42°$
3. $\angle MQN = 90°$ (inscribed in semi-circle; also $\angle MQN = \frac{1}{2}\overset{\frown}{MN}$)
4. $\angle MPN = 90°$ (inscribed in semi-circle; also $\angle MPN = \frac{1}{2}\overset{\frown}{MN}$)

EXERCISES 3-1-E

1. $\overset{\frown}{BC} = 2\angle BAC = 2(50°) = 100°$
2. $\overset{\frown}{AC} + \overset{\frown}{AB} + \overset{\frown}{BC} = 360°$; $\overset{\frown}{AB} = \overset{\frown}{AC}$; $\overset{\frown}{BC} = 100°$; $\overset{\frown}{AC} + \overset{\frown}{AC} + 100° = 360°$;
   $2\overset{\frown}{AC} = 260°$; $\overset{\frown}{AC} = 130°$

EXERCISES 3-1-F

1. $\angle ACO = 90°$ (tangent); $\angle OAC + \angle ACO + \angle AOC = 180°$;
   $30° + 90° + \angle AOC = 180°$; $\angle AOC = 60°$; $\angle AOB = \angle AOC = 60°$
2. $\angle BOC = \angle AOB + \angle AOC = 60° + 60° = 120°$; $\overset{\frown}{BC} = \angle BOC = 120°$

EXERCISES 3-1-Section

1. OB, OC, OG 3. AE, GD 5. GC, BC 7. $\overset{\frown}{BG}, \overset{\frown}{CG}$ 9. $\angle BOG$
11. AF and OG 13. $60°$ 15. $30°$ 17. $110°$ 19. $290°$
21. $90°$ 23. $35°$ 25. $93,000,000$ mi   27. $180°$

EXERCISES 3-2-A

1. (a) $c = \pi(8) = 8\pi$ ft;   (b) $c = 8\pi = 8(3.14) = 25.1$ ft
2. (a) $c = 2\pi(20) = 40\pi$ m;   (b) $c = 40\pi = 40(3.14) = 126$ m
3. $c = \pi(86.5) = (3.14)(86.5) = 272$ mm   4. $c = 2\pi(0.962) = 2(3.14)(0.962) = 6.04$ in.

EXERCISES 3-2-B

1. $L = \dfrac{60°\pi(29.0)}{180°} = \dfrac{(3.14)(29.0)}{3} = 30.4$ mm   2. $L = \dfrac{160°\pi(0.625)}{180°} = \dfrac{8(3.14)(0.625)}{9} = 1.74$ ft

3. $L = \dfrac{50°\pi(18.3)}{180°} = \dfrac{5(3.14)(18.3)}{18} = 16.0$ in.   4. $L = \dfrac{200°\pi(495)}{180°} = \dfrac{10(3.14)(495)}{9} = 1730$ m

EXERCISES 3-2-C

1. $20\pi = 2\pi r$; $r = 10$ ft   2. $36.0 = 2\pi r$; $r = 5.73$ m
3. $10\pi = \dfrac{45°\pi r}{180°}$; $10\pi = \dfrac{\pi r}{4}$; $r = 40$ mm   4. $1.08 = \dfrac{10°\pi r}{180°}$; $1.08 = \dfrac{3.14 r}{18}$; $r = 6.19$ mi

EXERCISES 3-2-D

1. $p = 2h + 2r + \pi r = 2(6.50) + 2(2.25) + (3.14)(2.25) = 13.0 + 4.50 + 7.06 = 24.6$ (rounded off)
2. $p = 2r + \dfrac{90°\pi r}{180°} = 2r + \dfrac{\pi r}{2} = 2(17.3) + \dfrac{(3.14)(17.3)}{2} = 34.6 + 27.2 = 61.8$ mm

EXERCISES 3-2-Section

1. (a) 10π ft   (b) 31.4 ft   3. (a) 10π mm   (b) 31.4 mm
5. 62.8 in.   7. 2.07 m   9. 542 yd   11. 18.3 km   13. 82.0 cm
15. 0.154 ft   17. 63.0 m   19. 80.5 in.   21. 7.68 in.   23. 8.68 cm
25. 87.9 in.   27. 6.66 mm   29. 2.07 ft   31. 2090 m

EXERCISES 3-3-A

1. $A = \pi(10.0)^2 = 100\pi = 100(3.14) = 314$ sq ft
2. $A = \pi(58.0)^2 = (3.14)(3364) = 10{,}600$ sq m
3. $A = \pi(92.0)^2 = (3.14)(8464) = 26{,}600$ sq mm
4. $A = \pi(0.364)^2 = (3.14)(0.132496) = 0.416$ sq mi

EXERCISES 3-3-B

1. $A = \dfrac{135°\pi(200)^2}{360°} = \dfrac{3(3.14)(40{,}000)}{8} = 47{,}100$ sq mm
2. $A = \dfrac{50°\pi(37.0)^2}{360°} = \dfrac{5(3.14)(1369)}{36} = 597$ sq in.
3. $A = \dfrac{90°\pi(75.0)^2}{360°} = \dfrac{(3.14)(5625)}{4} = 4420$ sq ft
4. $A = \dfrac{150°\pi(2.91)^2}{360°} = \dfrac{5(3.14)(8.4681)}{12} = 11.1$ sq m

EXERCISES 3-3-C

1. $25\pi = \pi r^2$; $r^2 = 25$; $r = \sqrt{25} = 5$ cm
2. $2.92 = \pi r^2$; $r^2 = \dfrac{2.92}{3.14} = 0.930$; $r = \sqrt{0.930} = 0.964$ in.
3. $54.0 = \dfrac{60°\pi r^2}{360°}$; $54.0 = \dfrac{3.14 r^2}{6}$; $r^2 = \dfrac{6(54.0)}{3.14} = 103$; $r = \sqrt{103} = 10.1$ ft
4. $882 = \dfrac{105°\pi r^2}{360°}$; $882 = \dfrac{7(3.14)r^2}{24}$; $r^2 = \dfrac{24(882)}{7(3.14)} = 963$; $r = \sqrt{963} = 31.0$ mm

EXERCISES 3-3-D

1. $A = (1.20)(2)(0.550) + \dfrac{1}{2}(3.14)(0.550)^2 = 1.32 + 0.474 = 1.79$ sq m
2. $A = 3.14(8.60)^2 - 3.14(1.60)^2 = 232 - 8.04 = 224$ sq in.

EXERCISES 3-3-Section

1. 5020 sq in.
3. 2.77 sq m
5. 4800 sq yd
7. 6760 sq cm
9. 4970 sq cm
11. 0.155 sq ft
13. 723 sq m
15. 1930 sq in.
17. 5.41 in.
19. 24.0 cm
21. 20.3 sq cm
23. 226 sq m
25. 3.91 ft
27. 154 sq m

Solutions 107

EXERCISES FOR UNIT THREE (Section 3-4)

1. 70°   3. 90°   5. 120°   7. 40°   9. 25°   11. 65°
13. 53.4 cm   15. 26.7 ft   17. 16.5 in.   19. 536 mm   21. 35.4 sq cm
23. 0.290 sq ft   25. 1860 sq in.   27. 5.39 sq m   29. 74.0 mm
31. 23.5 ft   33. 297 cm   35. 33.2 sq in.   37. 24,900 mi; 26,000 mi
39. 6.36 m   41. 67,800 lb   43. 11.1 sq m

EXERCISES 4-1-A

1. $V = (5)(2)(8) = 80$ cu ft
2. $V = (10)(6)(8) = 480$ cu cm
3. $V = (14.5)(10.3)(11.6) = 1730$ cu mm
4. $V = (26.5)(23.2)(15.6) = 9590$ cu in.

EXERCISES 4-1-B

1. $V = 4^3 = 64$ cu ft
2. $V = 10^3 = 1000$ cu cm
3. $V = (89.3)^3 = 712,000$ cu mm
4. $V = (16.2)^3 = 4250$ cu yd

EXERCISES 4-1-C

1. $V = (320)(14.0) = 4480$ cu in.
2. $V = (4900)(72.0) = 353,000$ cu m
3. $B = \frac{1}{2}(46.0)(26.0) = 598$ sq mm; $V = (598)(15.4) = 9210$ cu mm
4. $B = \frac{1}{2}(6.25)(8.34) = 26.1$ sq ft; $V = (26.1)(3.19) = 83.3$ cu ft

EXERCISES 4-1-D

1. $L = 2(20)(10) + 2(15)(10) = 400 + 300 = 700$ sq in.
2. $L = 2(260)(150) + 2(130)(150) = 78,000 + 39,000 = 117,000$ sq mm
3. $p = 2.73 + 3.20 + 3.28 = 9.21$ m; $L = (9.21)(2.35) = 21.6$ sq m
4. $p = 17.0 + 16.5 + 23.2 + 32.5 = 89.2$ ft; $L = (89.2)(12.2) = 1090$ sq ft

EXERCISES 4-1-E

1. $A = 2(20)(15) + 2(15)(10) + 2(20)(10) = 600 + 300 + 400 = 1300$ sq in.
2. $A = 2(260)(130) + 2(130)(150) + 2(260)(150) = 67,600 + 39,000 + 78,000 = 184,600$ sq mm
3. $A = 6(2.70)^2 = 6(7.29) = 43.7$ sq m
4. $A = 6(17.8)^2 = 6(317) = 1900$ sq ft

108   Solutions

EXERCISES 4-1-Section

1. 120 cu ft
3. 25,100 cu mm
5. 4100 cu in.
7. 178,000 cu cm
9. 6900 cu ft
11. 99,000 cu cm
13. 8.85 cu ft
15. 22,300 cu cm
17. (a) 112 sq cm
(b) 202 sq cm
19. (a) 795 sq ft
(b) 1170 sq ft
21. (a) 2.69 sq m
(b) 4.03 sq m
23. (a) 4280 sq in.
(b) 6420 sq in.
25. (a) 1060 sq mm
(b) 2620 sq mm
27. (a) 12,300 sq in.
(b) 18,900 sq in.
29. 2940 cu ft
31. 14.3 gm
33. 12,400 sq cm
35. 594 sq in.

EXERCISES 4-2-A

1. $V = (3.14)(4.00)^2 (5.00) = 251$ cu cm
2. $V = (3.14)(1.50)^2 (2.50) = 17.7$ cu ft
3. $V = (3.14)(20.5)^2 (15.0) = 19,800$ cu in.
4. $V = (3.14)(0.375)^2 (0.685) = 0.302$ cu m

EXERCISES 4-2-B

1. $L = 2(3.14)(6.00)(5.00) = 188$ cu in.
2. $L = 2(3.14)(16.0)(12.00) = 1210$ cu m
3. $L = 2(3.14)(25.5)(46.3) = 7410$ cu mm
4. $L = 2(3.14)(2.33)(6.28) = 91.9$ cu ft

EXERCISES 4-2-C

1. $A = 2(3.14)(6.00)^2 + 2(3.14)(6.00)(5.00) = 226 + 188 = 414$ cu in.
2. $A = 2(3.14)(16.0)^2 + 2(3.14)(16.0)(12.0) = 1610 + 1210 = 2820$ cu m
3. $A = 2(3.14)(25.5)^2 + 2(3.14)(25.5)(46.3) = 4080 + 7410 = 11,490$ cu mm
4. $A = 2(3.14)(2.33)^2 + 2(3.14)(2.33)(6.28) = 34.1 + 91.9 = 126$ cu ft

EXERCISES 4-2-D

1. $V = 3.14(24.0)(1.25^2 - 1.00^2) = 42.4$ cu in.; $w = (0.320)(42.4) = 13.6$ lb
2. $V = 3.14(20.0)(5.70^2 - 4.50^2) = 769$ cu cm

EXERCISES 4-2-Section

1. 18,800 cu cm
3. 2540 cu ft
5. 14,700,000 cu mm
7. 14,800 cu in.
9. (a) 226,000 sq mm
(b) 791,000 sq mm
11. (a) 124 sq in.
(b) 546 sq in.
13. (a) 641 sq cm
(b) 1550 sq cm
15. (a) 3360 sq ft
(b) 15,100 sq ft
17. 5430 cu in.
19. 576,000 cu mm
21. 123,000 bu
23. 175,000 gm
25. 40.0 sq in.
27. 10,200,000 cu cm

EXERCISES 4-3-A

1. $V = \frac{1}{3}(4.00)(12.0) = 16.0$ cu in.
2. $V = \frac{1}{3}(45.0)(22.0) = 330$ cu m
3. $V = \frac{1}{3}(9.00)^2(17.0) = 459$ cu mm
4. $V = \frac{1}{3}(36.0)^2(7.25) = 3130$ cu ft

EXERCISES 4-3-B

1. $L = \frac{1}{2}(56.0)(21.0) = 588$ sq m
2. $L = \frac{1}{2}(8.36)(4.14) = 17.3$ sq ft
3. $s^2 = (3.00)^2 + (4.00)^2 = 25.0$; $s = 5.00$ yd; $L = \frac{1}{2}(4)(6.00)(5.00) = 60.0$ sq yd.
4. $s^2 = (15.0)^2 + (8.00)^2 = 289$; $s = 17.0$ mm; $L = \frac{1}{2}(4)(30.0)(17.0) = 1020$ sq mm

EXERCISES 4-3-C

1. $V = \frac{1}{3}(3.14)(8.00)^2(6.00) = 402$ cu cm
2. $V = \frac{1}{3}(3.14)(1.60)^2(2.40) = 6.43$ cu ft
3. $V = \frac{1}{3}(3.14)(30.0)^2(18.0) = 17,000$ cu in.
4. $V = \frac{1}{3}(3.14)(0.875)^2(0.329) = 0.264$ cu m

EXERCISES 4-3-D

1. $L = (3.14)(6.00)(16.0) = 301$ sq ft
2. $L = (3.14)(36.0)(45.0) = 5090$ sq cm
3. $s^2 = r^2 + h^2 = (17.0)^2 + (6.00)^2 = 325$; $s = 18.0$ cm; $L = (3.14)(17.0)(18.0) = 961$ sq cm
4. $s^2 = r^2 + h^2 = (1.80)^2 + (3.50)^2 = 15.5$; $s = 3.94$ in; $L = (3.14)(1.80)(3.94) = 22.3$ sq in.

EXERCISES 4-3-Section

1. 80.0 cu m
3. 54,000 cu ft
5. 958 cu mm
7. 3.13 cu yd
9. 6000 sq mm
11. 76.5 sq yd
13. 240 sq m
15. 644 sq ft
17. 1470 cu ft
19. 13.8 cu m
21. 587 cu yd
23. 27,100 cu cm
25. 198 sq in.
27. 7.91 sq m
29. 424 sq ft.
31. 156,000 sq mm
33. 3,330,000 cu yd
35. 27.2 sq m
37. 36,700 lb
39. 155 sq cm

EXERCISES 4-4-A

1. $V = \frac{4}{3}(3.14)(6.00)^3 = 904$ cu m
2. $V = \frac{4}{3}(3.14)(15.0)^3 = 14,100$ cu ft
3. $V = \frac{4}{3}(3.14)(2.10)^3 = 38.8$ cu in.
4. $V = \frac{4}{3}(3.14)(43.0)^3 = 333,000$ cu cm

EXERCISES 4-4-B

1. $A = 4(3.14)(9.00)^2 = 1020$ sq m
2. $A = 4(3.14)(18.0)^2 = 4070$ sq ft
3. $A = 4(3.14)(42.0)^2 = 22{,}200$ sq ft
4. $A = 4(3.14)(0.430)^2 = 2.32$ sq cm

EXERCISES 4-4-C

1. $V = (3.14)(30.0)^2(80.0) + \frac{2}{3}(3.14)(30.0)^3 = 226{,}000 + 56{,}500 = 282{,}500$ cu ft
2. $A = 2(3.14)(30.0)(80.0) + 2(3.14)(30.0)^2 = 15{,}100 + 5650 = 20{,}750$ sq ft

EXERCISES 4-4-Section

1. 113 cu ft
3. 11.5 cu m
5. 1440 cu yd
7. 53.6 cu cm
9. 11,300 sq in.
11. 145 sq cm
13. 0.785 sq mi
15. 376,000 sq mm
17. 156 cu cm
19. 197,000,000 sq mi
21. 143,000 liters
23. 0.204 ft

EXERCISES FOR UNIT FOUR (Section 4-5)

1. 600 cu ft
3. 3.60 cu m
5. 42.9 cu yd
7. 6190 cu cm
9. 97,700 cu in.
11. 656,000 cu mm
13. 2750 cu ft
15. 436 cu m
17. 3990 cu ft
19. 24,400 cu cm
21. 900 sq ft
23. 14.4 sq m
25. 73.5 sq yd
27. 2320 sq cm
29. 5430 sq in.
31. 44,500 sq mm
33. 1360 sq ft
35. 264 sq m
37. 1230 sq ft
39. 4070 sq cm
41. 8.85 lb
43. 204 sq m
45. 11,400 gal
47. 4.88 sq m
49. 1.10 cu in.
51. 1020 sq cm
53. 12.2 cu in.
55. 45.3 sq cm

APPENDIX A

Significant Digits and Rounding Off

When we perform calculations on numbers, we must consider the accuracy of these numbers, since this affects the accuracy of the results obtained. Most of the numbers involved in technical and scientific work are *approximate*, having been arrived at through some process of measurement. However, certain other numbers are *exact*, having been arrived at through some definition or counting process. We can determine whether or not a number is approximate or exact if we know how the number was determined.

Example A. If we measure the length of a rope to be 15.3 ft, we know that the 15.3 is approximate. A more precise measuring device may cause us to determine the length as 15.28 ft. However, regardless of the method of measurement used, we shall not be able to determine this length exactly.

If a voltage shown on a voltmeter is read as 116 volts, the 116 is approximate. A more precise voltmeter may show the voltage as 115.7 volts. However, this voltage cannot be determined exactly.

Example B. If a computer counts the cards it has processed and prints this number as 837, this 837 is exact. We know the number of cards was not 836 or 838. Since 837 was determined through a counting process, it is exact.

When we say that 60 seconds = 1 minute, the 60 is exact, since this is a definition. By this definition there are exactly 60 seconds in one minute.

When we are writing approximate numbers we often have to include some zeros so that the decimal point will be properly located. However, except for these zeros, all other digits are considered to be *significant digits*. When we make computations with approximate numbers, we must know the number of significant digits. The following example illustrates how we determine this.

Example C. All numbers in this example are assumed to be approximate.

34.7 has three significant digits.

8900 has two significant digits. We assume that the two zeros are place holders (unless we have specific knowledge to the contrary.)

0.039 has two significant digits. The zeros are for proper location of the decimal point.

706.1 has four significant digits. The zero is not used for the location of the decimal point. It shows specifically the number of tens in the number.

5.90 has three significant digits. The zero is not necessary as a place holder, and should not be written unless it is significant.

Other approximate numbers with the proper number of significant digits are listed below.

96000	two	0.0709	three	1.070	four
30900	three	6.000	four	700.00	five
4.006	four	0.0005	one	20008	five

Note from the example above that all nonzero digits are significant. Zeros, other than those used as place holders for proper positioning of the decimal point, are also significant.

In computations involving approximate numbers, the position of the decimal point as well as the number of significant digits is important. The *precision* of a number refers directly to the decimal position of the last significant digit, whereas the *accuracy* of a number refers to the number of significant digits in the number. Consider the illustrations in the following example.

Example D. Suppose that you are measuring an electric current with two ammeters. One ammeter reads 0.031 amp and the second ammeter reads 0.0312 amp. The second reading is more precise, in that the last significant digit is the number of ten-thousandths, and the first reading is expressed only to thousandths. The second reading is also more accurate, since it has three significant digits rather than two.

A machine part is measured to be 2.5 cm long. It is coated with a film 0.025 cm thick. The thickness of the film has been measured to a greater precision, although the two measurements have the same accuracy: two significant digits.

A segment of a newly completed highway is 9270 ft long. The concrete surface is 0.8 ft thick. Of these two numbers, 9270 is more accurate, since it contains three significant digits, and 0.8 is more precise, since it is expressed to tenths.

The last significant digit of an approximate number is known not to be completely accurate. It has usually been determined by estimation or *rounding off*. However, we do know that it is at most in error by one-half of a unit in its place value.

Example E. When we measure the length of the rope referred to in Example A to be 15.3 ft, we are saying that the length is at least 15.25 ft and no longer than 15.35 ft. Any value between these two, rounded off to tenths, would be expressed as 15.3 ft.

In converting the fraction $\frac{2}{3}$ to the decimal form 0.667, we are saying that the value is between 0.6665 and 0.6675.

The principle of rounding off a number is to write the closest approximation, with the last significant digit in a specified position, or with a specified number of significant digits. We shall now formalize the process of rounding off as follows: If we want a certain number of significant digits, we examine the digit in the next place to the right. If this digit is less than 5, we accept the digit in the last place. If the next digit is 5 or greater, we increase the digit in the last place by 1, and this resulting digit becomes the final significant digit of the approximation. If necessary, we use zeros to replace other digits in order to locate the decimal point properly. Except when the next digit is a 5, and no other nonzero digits are discarded, we have the closest possible approximation with the desired number of significant digits.

Example F.   70360 rounded off to three significant digits is 70400.
70430 rounded off to three significant digits is 70400.
187.35 rounded off to four significant digits is 187.4.
71500 rounded off to two significant digits is 72000.

APPENDIX B
Table of Squares and Square Roots

n	n^2	$\sqrt{n}$	$\sqrt{10n}$	n	n^2	$\sqrt{n}$	$\sqrt{10n}$	n	n^2	$\sqrt{n}$	$\sqrt{10n}$
1.00	1.0000	1.00000	3.16228	**1.50**	2.2500	1.22474	3.87298	**2.00**	4.0000	1.41421	4.47214
1.01	1.0201	1.00499	3.17805	1.51	2.2801	1.22882	3.88587	2.01	4.0401	1.41774	4.48330
1.02	1.0404	1.00995	3.19374	1.52	2.3104	1.23288	3.89872	2.02	4.0804	1.42127	4.49444
1.03	1.0609	1.01489	3.20936	1.53	2.3409	1.23693	3.91152	2.03	4.1209	1.42478	4.50555
1.04	1.0816	1.01980	3.22490	1.54	2.3716	1.24097	3.92428	2.04	4.1616	1.42829	4.51664
1.05	1.1025	1.02470	3.24037	1.55	2.4025	1.24499	3.93700	2.05	4.2025	1.43178	4.52769
1.06	1.1236	1.02956	3.25576	1.56	2.4336	1.24900	3.94968	2.06	4.2436	1.43527	4.53872
1.07	1.1449	1.03441	3.27109	1.57	2.4649	1.25300	3.96232	2.07	4.2849	1.43875	4.54973
1.08	1.1664	1.03923	3.28634	1.58	2.4964	1.25698	3.97492	2.08	4.3264	1.44222	4.56070
1.09	1.1881	1.04403	3.30151	1.59	2.5281	1.26095	3.98748	2.09	4.3681	1.44568	4.57165
1.10	1.2100	1.04881	3.31662	**1.60**	2.5600	1.26491	4.00000	**2.10**	4.4100	1.44914	4.58258
1.11	1.2321	1.05357	3.33167	1.61	2.5921	1.26886	4.01248	2.11	4.4521	1.45258	4.59347
1.12	1.2544	1.05830	3.34664	1.62	2.6244	1.27279	4.02492	2.12	4.4944	1.45602	4.60435
1.13	1.2769	1.06301	3.36155	1.63	2.6569	1.27671	4.03733	2.13	4.5369	1.45945	4.61519
1.14	1.2996	1.06771	3.37639	1.64	2.6896	1.28062	4.04969	2.14	4.5796	1.46287	4.62601
1.15	1.3225	1.07238	3.39116	1.65	2.7225	1.28452	4.06202	2.15	4.6225	1.46629	4.63681
1.16	1.3456	1.07703	3.40588	1.66	2.7556	1.28841	4.07431	2.16	4.6656	1.46969	4.64758
1.17	1.3689	1.08167	3.42053	1.67	2.7889	1.29228	4.08656	2.17	4.7089	1.47309	4.65833
1.18	1.3924	1.08628	3.43511	1.68	2.8224	1.29615	4.09878	2.18	4.7524	1.47648	4.66905
1.19	1.4161	1.09087	3.44964	1.69	2.8561	1.30000	4.11096	2.19	4.7961	1.47986	4.67974
1.20	1.4400	1.09545	3.46410	**1.70**	2.8900	1.30384	4.12311	**2.20**	4.8400	1.48324	4.69042
1.21	1.4641	1.10000	3.47851	1.71	2.9241	1.30767	4.13521	2.21	4.8841	1.48661	4.70106
1.22	1.4884	1.10454	3.49285	1.72	2.9584	1.31149	4.14729	2.22	4.9284	1.48997	4.71169
1.23	1.5129	1.10905	3.50714	1.73	2.9929	1.31529	4.15933	2.23	4.9729	1.49332	4.72229
1.24	1.5376	1.11355	3.52136	1.74	3.0276	1.31909	4.17133	2.24	5.0176	1.49666	4.73286
1.25	1.5625	1.11803	3.53553	1.75	3.0625	1.32288	4.18330	2.25	5.0625	1.50000	4.74342
1.26	1.5876	1.12250	3.54965	1.76	3.0976	1.32665	4.19524	2.26	5.1076	1.50333	4.75395
1.27	1.6129	1.12694	3.56371	1.77	3.1329	1.33041	4.20714	2.27	5.1529	1.50665	4.76445
1.28	1.6384	1.13137	3.57771	1.78	3.1684	1.33417	4.21900	2.28	5.1984	1.50997	4.77493
1.29	1.6641	1.13578	3.59166	1.79	3.2041	1.33791	4.23084	2.29	5.2441	1.51327	4.78539
1.30	1.6900	1.14018	3.60555	**1.80**	3.2400	1.34164	4.24264	**2.30**	5.2900	1.51658	4.79583
1.31	1.7161	1.14455	3.61939	1.81	3.2761	1.34536	4.25441	2.31	5.3361	1.51987	4.80625
1.32	1.7424	1.14891	3.63318	1.82	3.3124	1.34907	4.26615	2.32	5.3824	1.52315	4.81664
1.33	1.7689	1.15326	3.64692	1.83	3.3489	1.35277	4.27785	2.33	5.4289	1.52643	4.82701
1.34	1.7956	1.15758	3.66060	1.84	3.3856	1.35647	4.28952	2.34	5.4756	1.52971	4.83735
1.35	1.8225	1.16190	3.67423	1.85	3.4225	1.36015	4.30116	2.35	5.5225	1.53297	4.84768
1.36	1.8496	1.16619	3.68782	1.86	3.4596	1.36382	4.31277	2.36	5.5696	1.53623	4.85798
1.37	1.8769	1.17047	3.70135	1.87	3.4969	1.36748	4.32435	2.37	5.6169	1.53948	4.86826
1.38	1.9044	1.17473	3.71484	1.88	3.5344	1.37113	4.33590	2.38	5.6644	1.54272	4.87852
1.39	1.9321	1.17898	3.72827	1.89	3.5721	1.37477	4.34741	2.39	5.7121	1.54596	4.88876
1.40	1.9600	1.18322	3.74166	**1.90**	3.6100	1.37840	4.35890	**2.40**	5.7600	1.54919	4.89898
1.41	1.9881	1.18743	3.75500	1.91	3.6481	1.38203	4.37035	2.41	5.8081	1.55242	4.90918
1.42	2.0164	1.19164	3.76829	1.92	3.6864	1.38564	4.38178	2.42	5.8564	1.55563	4.91935
1.43	2.0449	1.19583	3.78153	1.93	3.7249	1.38924	4.39318	2.43	5.9049	1.55885	4.92950
1.44	2.0736	1.20000	3.79473	1.94	3.7636	1.39284	4.40454	2.44	5.9536	1.56205	4.93964
1.45	2.1025	1.20416	3.80789	1.95	3.8025	1.39642	4.41588	2.45	6.0025	1.56525	4.94975
1.46	2.1316	1.20830	3.82099	1.96	3.8416	1.40000	4.42719	2.46	6.0516	1.56844	4.95984
1.47	2.1609	1.21244	3.83406	1.97	3.8809	1.40357	4.43847	2.47	6.1009	1.57162	4.96991
1.48	2.1904	1.21655	3.84708	1.98	3.9204	1.40712	4.44972	2.48	6.1504	1.57480	4.97996
1.49	2.2201	1.22066	3.86005	1.99	3.9601	1.41067	4.46094	2.49	6.2001	1.57797	4.98999
1.50	2.2500	1.22474	3.87298	**2.00**	4.0000	1.41421	4.47214	**2.50**	6.2500	1.58114	5.00000
n	n^2	$\sqrt{n}$	$\sqrt{10n}$	n	n^2	$\sqrt{n}$	$\sqrt{10n}$	n	n^2	$\sqrt{n}$	$\sqrt{10n}$

Illustrations: for $n = 1.27$

$1.27^2 = 1.6129$ $0.127^2 = 0.016129$

$12.7^2 = 161.29$ $0.0127^2 = 0.00016129$

$127^2 = 16129$ $0.00127^2 = 0.0000016129$

$1270^2 = 1612900$

Table of Squares and Square Roots *(Continued)*

n	n²	√n	√10n	n	n²	√n	√10n	n	n²	√n	√10n
2.50	6.2500	1.58114	5.00000	**3.00**	9.0000	1.73205	5.47723	**3.50**	12.2500	1.87083	5.91608
2.51	6.3001	1.58430	5.00999	3.01	9.0601	1.73494	5.48635	3.51	12.3201	1.87350	5.92453
2.52	6.3504	1.58745	5.01996	3.02	9.1204	1.73781	5.49545	3.52	12.3904	1.87617	5.93296
2.53	6.4009	1.59060	5.02991	3.03	9.1809	1.74069	5.50454	3.53	12.4609	1.87883	5.94138
2.54	6.4516	1.59374	5.03984	3.04	9.2416	1.74356	5.51362	3.54	12.5316	1.88149	5.94979
2.55	6.5025	1.59687	5.04975	3.05	9.3025	1.74642	5.52268	3.55	12.6025	1.88414	5.95819
2.56	6.5536	1.60000	5.05964	3.06	9.3636	1.74929	5.53173	3.56	12.6736	1.88680	5.96657
2.57	6.6049	1.60312	5.06952	3.07	9.4249	1.75214	5.54076	3.57	12.7449	1.88944	5.97495
2.58	6.6564	1.60624	5.07937	3.08	9.4864	1.75499	5.54977	3.58	12.8164	1.89209	5.98331
2.59	6.7081	1.60935	5.08920	3.09	9.5481	1.75784	5.55878	3.59	12.8881	1.89473	5.99166
2.60	6.7600	1.61245	5.09902	**3.10**	9.6100	1.76068	5.56776	**3.60**	12.9600	1.89737	6.00000
2.61	6.8121	1.61555	5.10882	3.11	9.6721	1.76352	5.57674	3.61	13.0321	1.90000	6.00833
2.62	6.8644	1.61864	5.11859	3.12	9.7344	1.76635	5.58570	3.62	13.1044	1.90263	6.01664
2.63	6.9169	1.62173	5.12835	3.13	9.7969	1.76918	5.59464	3.63	13.1769	1.90526	6.02495
2.64	6.9696	1.62481	5.13809	3.14	9.8596	1.77200	5.60357	3.64	13.2496	1.90788	6.03324
2.65	7.0225	1.62788	5.14782	3.15	9.9225	1.77482	5.61249	3.65	13.3225	1.91050	6.04152
2.66	7.0756	1.63095	5.15752	3.16	9.9856	1.77764	5.62139	3.66	13.3956	1.91311	6.04979
2.67	7.1289	1.63401	5.16720	3.17	10.0489	1.78045	5.63028	3.67	13.4689	1.91572	6.05805
2.68	7.1824	1.63707	5.17687	3.18	10.1124	1.78326	5.63915	3.68	13.5424	1.91833	6.06630
2.69	7.2361	1.64012	5.18652	3.19	10.1761	1.78606	5.64801	3.69	13.6161	1.92094	6.07454
2.70	7.2900	1.64317	5.19615	**3.20**	10.2400	1.78885	5.65685	**3.70**	13.6900	1.92354	6.08276
2.71	7.3441	1.64621	5.20577	3.21	10.3041	1.79165	5.66569	3.71	13.7641	1.92614	6.09098
2.72	7.3984	1.64924	5.21536	3.22	10.3684	1.79444	5.67450	3.72	13.8384	1.92873	6.09918
2.73	7.4529	1.65227	5.22494	3.23	10.4329	1.79722	5.68331	3.73	13.9129	1.93132	6.10737
2.74	7.5076	1.65529	5.23450	3.24	10.4976	1.80000	5.69210	3.74	13.9876	1.93391	6.11555
2.75	7.5625	1.65831	5.24404	3.25	10.5625	1.80278	5.70088	3.75	14.0625	1.93649	6.12372
2.76	7.6176	1.66132	5.25357	3.26	10.6276	1.80555	5.70964	3.76	14.1376	1.93907	6.13188
2.77	7.6729	1.66433	5.26308	3.27	10.6929	1.80831	5.71839	3.77	14.2129	1.94165	6.14003
2.78	7.7284	1.66733	5.27257	3.28	10.7584	1.81108	5.72713	3.78	14.2884	1.94422	6.14817
2.79	7.7841	1.67033	5.28205	3.29	10.8241	1.81384	5.73585	3.79	14.3641	1.94679	6.15630
2.80	7.8400	1.67332	5.29150	**3.30**	10.8900	1.81659	5.74456	**3.80**	14.4400	1.94936	6.16441
2.81	7.8961	1.67631	5.30094	3.31	10.9561	1.81934	5.75326	3.81	14.5161	1.95192	6.17252
2.82	7.9524	1.67929	5.31037	3.32	11.0224	1.82209	5.76194	3.82	14.5924	1.95448	6.18061
2.83	8.0089	1.68226	5.31977	3.33	11.0889	1.82483	5.77062	3.83	14.6689	1.95704	6.18870
2.84	8.0656	1.68523	5.32917	3.34	11.1556	1.82757	5.77927	3.84	14.7456	1.95959	6.19677
2.85	8.1225	1.68819	5.33854	3.35	11.2225	1.83030	5.78792	3.85	14.8225	1.96214	6.20484
2.86	8.1796	1.69115	5.34790	3.36	11.2896	1.83303	5.79655	3.86	14.8996	1.96469	6.21289
2.87	8.2369	1.69411	5.35724	3.37	11.3569	1.83576	5.80517	3.87	14.9769	1.96723	6.22093
2.88	8.2944	1.69706	5.36656	3.38	11.4244	1.83848	5.81378	3.88	15.0544	1.96977	6.22896
2.89	8.3521	1.70000	5.37587	3.39	11.4921	1.84120	5.82237	3.89	15.1321	1.97231	6.23699
2.90	8.4100	1.70294	5.38516	**3.40**	11.5600	1.84391	5.83095	**3.90**	15.2100	1.97484	6.24500
2.91	8.4681	1.70587	5.39444	3.41	11.6281	1.84662	5.83952	3.91	15.2881	1.97737	6.25300
2.92	8.5264	1.70880	5.40370	3.42	11.6964	1.84932	5.84808	3.92	15.3664	1.97990	6.26099
2.93	8.5849	1.71172	5.41295	3.43	11.7649	1.85203	5.85662	3.93	15.4449	1.98242	6.26897
2.94	8.6436	1.71464	5.42218	3.44	11.8336	1.85472	5.86515	3.94	15.5236	1.98494	6.27694
2.95	8.7025	1.71756	5.43139	3.45	11.9025	1.85742	5.87367	3.95	15.6025	1.98746	6.28490
2.96	8.7616	1.72047	5.44059	3.46	11.9716	1.86011	5.88218	3.96	15.6816	1.98997	6.29285
2.97	8.8209	1.72337	5.44977	3.47	12.0409	1.86279	5.89067	3.97	15.7609	1.99249	6.30079
2.98	8.8804	1.72627	5.45894	3.48	12.1104	1.86548	5.89915	3.98	15.8404	1.99499	6.30872
2.99	8.9401	1.72916	5.46809	3.49	12.1801	1.86815	5.90762	3.99	15.9201	1.99750	6.31664
3.00	9.0000	1.73205	5.47723	**3.50**	12.2500	1.87083	5.91608	**4.00**	16.0000	2.00000	6.32456

Illustrations: for $n = 1.27$

$\sqrt{1.27} = 1.12694$ $\qquad$ $\sqrt{0.127} = 0.356371$

$\sqrt{12.7} = 3.56371$ $\qquad$ $\sqrt{0.0127} = 0.112694$

$\sqrt{127} = 11.2694$ $\qquad$ $\sqrt{0.00127} = 0.0356371$

$\sqrt{1270} = 35.6371$ $\qquad$ $\sqrt{0.000127} = 0.0112694$

Table of Squares and Square Roots *(Continued)*

n	n²	√n	√10n	n	n²	√n	√10n	n	n²	√n	√10n
4.00	16.0000	2.00000	6.32456	**4.50**	20.2500	2.12132	6.70820	**5.00**	25.0000	2.23607	7.07107
4.01	16.0801	2.00250	6.33246	4.51	20.3401	2.12368	6.71565	5.01	25.1001	2.23830	7.07814
4.02	16.1604	2.00499	6.34035	4.52	20.4304	2.12603	6.72309	5.02	25.2004	2.24054	7.08520
4.03	16.2409	2.00749	6.34823	4.53	20.5209	2.12838	6.73053	5.03	25.3009	2.24277	7.09225
4.04	16.3216	2.00998	6.35610	4.54	20.6116	2.13073	6.73795	5.04	25.4016	2.24499	7.09930
4.05	16.4025	2.01246	6.36396	4.55	20.7025	2.13307	6.74537	5.05	25.5025	2.24722	7.10634
4.06	16.4836	2.01494	6.37181	4.56	20.7936	2.13542	6.75278	5.06	25.6036	2.24944	7.11337
4.07	16.5649	2.01742	6.37966	4.57	20.8849	2.13776	6.76018	5.07	25.7049	2.25167	7.12039
4.08	16.6464	2.01990	6.38749	4.58	20.9764	2.14009	6.76757	5.08	25.8064	2.25389	7.12741
4.09	16.7281	2.02237	6.39531	4.59	21.0681	2.14243	6.77495	5.09	25.9081	2.25610	7.13442
4.10	16.8100	2.02485	6.40312	**4.60**	21.1600	2.14476	6.78233	**5.10**	26.0100	2.25832	7.14143
4.11	16.8921	2.02731	6.41093	4.61	21.2521	2.14709	6.78970	5.11	26.1121	2.26053	7.14843
4.12	16.9744	2.02978	6.41872	4.62	21.3444	2.14942	6.79706	5.12	26.2144	2.26274	7.15542
4.13	17.0569	2.03224	6.42651	4.63	21.4369	2.15174	6.80441	5.13	26.3169	2.26495	7.16240
4.14	17.1396	2.03470	6.43428	4.64	21.5296	2.15407	6.81175	5.14	26.4196	2.26716	7.16938
4.15	17.2225	2.03715	6.44205	4.65	21.6225	2.15639	6.81909	5.15	26.5225	2.26936	7.17635
4.16	17.3056	2.03961	6.44981	4.66	21.7156	2.15870	6.82642	5.16	26.6256	2.27156	7.18331
4.17	17.3889	2.04206	6.45755	4.67	21.8089	2.16102	6.83374	5.17	26.7289	2.27376	7.19027
4.18	17.4724	2.04450	6.46529	4.68	21.9024	2.16333	6.84105	5.18	26.8324	2.27596	7.19722
4.19	17.5561	2.04695	6.47302	4.69	21.9961	2.16564	6.84836	5.19	26.9361	2.27816	7.20417
4.20	17.6400	2.04939	6.48074	**4.70**	22.0900	2.16795	6.85565	**5.20**	27.0400	2.28035	7.21110
4.21	17.7241	2.05183	6.48845	4.71	22.1841	2.17025	6.86294	5.21	27.1441	2.28254	7.21803
4.22	17.8084	2.05426	6.49615	4.72	22.2784	2.17256	6.87023	5.22	27.2484	2.28473	7.22496
4.23	17.8929	2.05670	6.50384	4.73	22.3729	2.17486	6.87750	5.23	27.3529	2.28692	7.23187
4.24	17.9776	2.05913	6.51153	4.74	22.4676	2.17715	6.88477	5.24	27.4576	2.28910	7.23878
4.25	18.0625	2.06155	6.51920	4.75	22.5625	2.17945	6.89202	5.25	27.5625	2.29129	7.24569
4.26	18.1476	2.06398	6.52687	4.76	22.6576	2.18174	6.89928	5.26	27.6676	2.29347	7.25259
4.27	18.2329	2.06640	6.53452	4.77	22.7529	2.18403	6.90652	5.27	27.7729	2.29565	7.25948
4.28	18.3184	2.06882	6.54217	4.78	22.8484	2.18632	6.91375	5.28	27.8784	2.29783	7.26636
4.29	18.4041	2.07123	6.54981	4.79	22.9441	2.18861	6.92098	5.29	27.9841	2.30000	7.27324
4.30	18.4900	2.07364	6.55744	**4.80**	23.0400	2.19089	6.92820	**5.30**	28.0900	2.30217	7.28011
4.31	18.5761	2.07605	6.56506	4.81	23.1361	2.19317	6.93542	5.31	28.1961	2.30434	7.28697
4.32	18.6624	2.07846	6.57267	4.82	23.2324	2.19545	6.94262	5.32	28.3024	2.30651	7.29383
4.33	18.7489	2.08087	6.58027	4.83	23.3289	2.19773	6.94982	5.33	28.4089	2.30868	7.30068
4.34	18.8356	2.08327	6.58787	4.84	23.4256	2.20000	6.95701	5.34	28.5156	2.31084	7.30753
4.35	18.9225	2.08567	6.59545	4.85	23.5225	2.20227	6.96419	5.35	28.6225	2.31301	7.31437
4.36	19.0096	2.08806	6.60303	4.86	23.6196	2.20454	6.97137	5.36	28.7296	2.31517	7.32120
4.37	19.0969	2.09045	6.61060	4.87	23.7169	2.20681	6.97854	5.37	28.8369	2.31733	7.32803
4.38	19.1844	2.09284	6.61816	4.88	23.8144	2.20907	6.98570	5.38	28.9444	2.31948	7.33485
4.39	19.2721	2.09523	6.62571	4.89	23.9121	2.21133	6.99285	5.39	29.0521	2.32164	7.34166
4.40	19.3600	2.09762	6.63325	**4.90**	24.0100	2.21359	7.00000	**5.40**	29.1600	2.32379	7.34847
4.41	19.4481	2.10000	6.64078	4.91	24.1081	2.21585	7.00714	5.41	29.2681	2.32594	7.35527
4.42	19.5364	2.10238	6.64831	4.92	24.2064	2.21811	7.01427	5.42	29.3764	2.32809	7.36206
4.43	19.6249	2.10476	6.65582	4.93	24.3049	2.22036	7.02140	5.43	29.4849	2.33024	7.36885
4.44	19.7136	2.10713	6.66333	4.94	24.4036	2.22261	7.02851	5.44	29.5936	2.33238	7.37564
4.45	19.8025	2.10950	6.67083	4.95	24.5025	2.22486	7.03562	5.45	29.7025	2.33452	7.38241
4.46	19.8916	2.11187	6.67832	4.96	24.6016	2.22711	7.04273	5.46	29.8116	2.33666	7.38918
4.47	19.9809	2.11424	6.68581	4.97	24.7009	2.22935	7.04982	5.47	29.9209	2.33880	7.39594
4.48	20.0704	2.11660	6.69328	4.98	24.8004	2.23159	7.05691	5.48	30.0304	2.34094	7.40270
4.49	20.1601	2.11896	6.70075	4.99	24.9001	2.23383	7.06399	5.49	30.1401	2.34307	7.40945
4.50	20.2500	2.12132	6.70820	**5.00**	25.0000	2.23607	7.07107	**5.50**	30.2500	2.34521	7.41620

Illustrations: for $n = 4.61$

$4.61^2 = 21.2521$ $0.461^2 = 0.212521$

$46.1^2 = 2125.21$ $0.0461^2 = 0.00212521$

$461^2 = 212521$ $0.00461^2 = 0.0000212521$

$4610^2 = 21252100$

Appendix B

Table of Squares and Square Roots *(Continued)*

n	n^2	$\sqrt{n}$	$\sqrt{10n}$	n	n^2	$\sqrt{n}$	$\sqrt{10n}$	n	n^2	$\sqrt{n}$	$\sqrt{10n}$
5.50	30.2500	2.34521	7.41620	6.00	36.0000	2.44949	7.74597	6.50	42.2500	2.54951	8.06226
5.51	30.3601	2.34734	7.42294	6.01	36.1201	2.45153	7.75242	6.51	42.3801	2.55147	8.06846
5.52	30.4704	2.34947	7.42967	6.02	36.2404	2.45357	7.75887	6.52	42.5104	2.55343	8.07465
5.53	30.5809	2.35160	7.43640	6.03	36.3609	2.45561	7.76531	6.53	42.6409	2.55539	8.08084
5.54	30.6916	2.35372	7.44312	6.04	36.4816	2.45764	7.77174	6.54	42.7716	2.55734	8.08703
5.55	30.8025	2.35584	7.44983	6.05	36.6025	2.45967	7.77817	6.55	42.9025	2.55930	8.09321
5.56	30.9136	2.35797	7.45654	6.06	36.7236	2.46171	7.78460	6.56	43.0336	2.56125	8.09938
5.57	31.0249	2.36008	7.46324	6.07	36.8449	2.46374	7.79102	6.57	43.1649	2.56320	8.10555
5.58	31.1364	2.36220	7.46994	6.08	36.9664	2.46577	7.79744	6.58	43.2964	2.56515	8.11172
5.59	31.2481	2.36432	7.47663	6.09	37.0881	2.46779	7.80385	6.59	43.4281	2.56710	8.11788
5.60	31.3600	2.36643	7.48331	6.10	37.2100	2.46982	7.81025	6.60	43.5600	2.56905	8.12404
5.61	31.4721	2.26854	7.48999	6.11	37.3321	2.47184	7.81665	6.61	43.6921	2.57099	8.13019
5.62	31.5844	2.37065	7.49667	6.12	37.4544	2.47386	7.82304	6.62	43.8244	2.57294	8.13634
5.63	31.6969	2.37276	7.50333	6.13	37.5769	2.47588	7.82943	6.63	43.9569	2.57488	8.14248
5.64	31.8096	2.37487	7.50999	6.14	37.6996	2.47790	7.83582	6.64	44.0896	2.57682	8.14862
5.65	31.9225	2.37697	7.51665	6.15	37.8225	2.47992	7.84219	6.65	44.2225	2.57876	8.15475
5.66	32.0356	2.37908	7.52330	6.16	37.9456	2.48193	7.84857	6.66	44.3556	2.58070	8.16088
5.67	32.1489	2.38118	7.52994	6.17	38.0689	2.48395	7.85493	6.67	44.4889	2.58263	8.16701
5.68	32.2624	2.38328	7.53658	6.18	38.1924	2.48596	7.86130	6.68	44.6224	2.58457	8.17313
5.69	32.3761	2.38537	7.54321	6.19	38.3161	2.48797	7.86766	6.69	44.7561	2.58650	8.17924
5.70	32.4900	2.38747	7.54983	6.20	38.4400	2.48998	7.87401	6.70	44.8900	2.58844	8.18535
5.71	32.6041	2.38956	7.55645	6.21	38.5641	2.49199	7.88036	6.71	45.0241	2.59037	8.19146
5.72	32.7184	2.39165	7.56307	6.22	38.6884	2.49399	7.88670	6.72	45.1584	2.59230	8.19756
5.73	32.8329	2.39374	7.56968	6.23	38.8129	2.49600	7.89303	6.73	45.2929	2.59422	8.20366
5.74	32.9476	2.39583	7.57628	6.24	38.9376	2.49800	7.89937	6.74	45.4276	2.59615	8.20975
5.75	33.0625	2.39792	7.58288	6.25	39.0625	2.50000	7.90569	6.75	45.5625	2.59808	8.21584
5.76	33.1776	2.40000	7.58947	6.26	39.1876	2.50200	7.91202	6.76	45.6976	2.60000	8.22192
5.77	33.2929	2.40208	7.59605	6.27	39.3129	2.50400	7.91833	6.77	45.8329	2.60192	8.22800
5.78	33.4084	2.40416	7.60263	6.28	39.4384	2.50599	7.92465	6.78	45.9684	2.60384	8.23408
5.79	33.5241	2.40624	7.60920	6.29	39.5641	2.50799	7.93095	6.79	46.1041	2.60576	8.24015
5.80	33.6400	2.40832	7.61577	6.30	39.6900	2.50998	7.93725	6.80	46.2400	2.60768	8.24621
5.81	33.7561	2.41039	7.62234	6.31	39.8161	2.51197	7.94355	6.81	46.3761	2.60960	8.25227
5.82	33.8724	2.41247	7.62889	6.32	39.9424	2.51396	7.94984	6.82	46.5124	2.61151	8.25833
5.83	33.9889	2.41454	7.63544	6.33	40.0689	2.51595	7.95613	6.83	46.6489	2.61343	8.26438
5.84	34.1056	2.41661	7.64199	6.34	40.1956	2.51794	7.96241	6.84	46.7856	2.61534	8.27043
5.85	34.2225	2.41868	7.64853	6.35	40.3225	2.51992	7.96869	6.85	46.9225	2.61725	8.27647
5.86	34.3396	2.42074	7.65506	6.36	40.4496	2.52190	7.97496	6.86	47.0596	2.61916	8.28251
5.87	34.4569	2.42281	7.66159	6.37	40.5769	2.52389	7.98123	6.87	47.1969	2.62107	8.28855
5.88	34.5744	2.42487	7.66812	6.38	40.7044	2.52587	7.98749	6.88	47.3344	2.62298	8.29458
5.89	34.6921	2.42693	7.67463	6.39	40.8321	2.52784	7.99375	6.89	47.4721	2.62488	8.30060
5.90	34.8100	2.42899	7.68115	6.40	40.9600	2.52982	8.00000	6.90	47.6100	2.62679	8.30662
5.91	34.9281	2.43105	7.68765	6.41	41.0881	2.53180	8.00625	6.91	47.7481	2.62869	8.31264
5.92	35.0464	2.43311	7.69415	6.42	41.2164	2.53377	8.01249	6.92	47.8864	2.63059	8.31865
5.93	35.1649	2.43516	7.70065	6.43	41.3449	2.53574	8.01873	6.93	48.0249	2.63249	8.32466
5.94	35.2836	2.43721	7.70714	6.44	41.4736	2.53772	8.02496	6.94	48.1636	2.63439	8.33067
5.95	35.4025	2.43926	7.71362	6.45	41.6025	2.53969	8.03119	6.95	48.3025	2.63629	8.33667
5.96	35.5216	2.44131	7.72010	6.46	41.7316	2.54165	8.03741	6.96	48.4416	2.63818	8.34266
5.97	35.6409	2.44336	7.72658	6.47	41.8609	2.54362	8.04363	6.97	48.5809	2.64008	8.34865
5.98	35.7604	2.44540	7.73305	6.48	41.9904	2.54558	8.04984	6.98	48.7204	2.64197	8.35464
5.99	35.8801	2.44745	7.73951	6.49	42.1201	2.54755	8.05605	6.99	48.8601	2.64386	8.36062
6.00	36.0000	2.44949	7.74597	6.50	42.2500	2.54951	8.06226	7.00	49.0000	2.64575	8.36660

Illustrations: for $n = 4.61$

$$\sqrt{4.61} = 2.14709 \qquad \sqrt{0.461} = 0.678970$$

$$\sqrt{46.1} = 6.78970 \qquad \sqrt{0.0461} = 0.214709$$

$$\sqrt{461} = 21.4709 \qquad \sqrt{0.00461} = 0.0678970$$

$$\sqrt{4610} = 67.8970 \qquad \sqrt{0.000461} = 0.0214709$$

Table of Squares and Square Roots *(Continued)*

n	n^2	$\sqrt{n}$	$\sqrt{10n}$	n	n^2	$\sqrt{n}$	$\sqrt{10n}$	n	n^2	$\sqrt{n}$	$\sqrt{10n}$
7.00	49.0000	2.64575	8.36660	7.50	56.2500	2.73861	8.66025	8.00	64.0000	2.82843	8.94427
7.01	49.1401	2.64764	8.37257	7.51	56.4001	2.74044	8.66603	8.01	64.1601	2.83019	8.94986
7.02	49.2804	2.64953	8.37854	7.52	56.5504	2.74226	8.67179	8.02	64.3204	2.83196	8.95545
7.03	49.4209	2.65141	8.38451	7.53	56.7009	2.74408	8.67756	8.03	64.4809	2.83373	8.96103
7.04	49.5616	2.65330	8.39047	7.54	56.8516	2.74591	8.68332	8.04	64.6416	2.83549	8.96660
7.05	49.7025	2.65518	8.39643	7.55	57.0025	2.74773	8.68907	8.05	64.8025	2.83725	8.97218
7.06	49.8436	2.65707	8.40238	7.56	57.1536	2.74955	8.69483	8.06	64.9636	2.83901	8.97775
7.07	49.9849	2.65895	8.40833	7.57	57.3049	2.75136	8.70057	8.07	65.1249	2.84077	8.98332
7.08	50.1264	2.66083	8.41427	7.58	57.4564	2.75318	8.70632	8.08	65.2864	2.84253	8.98888
7.09	50.2681	2.66271	8.42021	7.59	57.6081	2.75500	8.71206	8.09	65.4481	2.84429	8.99444
7.10	50.4100	2.66458	8.42615	7.60	57.7600	2.75681	8.71780	8.10	65.6100	2.84605	9.00000
7.11	50.5521	2.66646	8.43208	7.61	57.9121	2.75862	8.72353	8.11	65.7721	2.84781	9.00555
7.12	50.6944	2.66833	8.43801	7.62	58.0644	2.76043	8.72926	8.12	65.9344	2.84956	9.01110
7.13	50.8369	2.67021	8.44393	7.63	58.2169	2.76225	8.73499	8.13	66.0969	2.85132	9.01665
7.14	50.9796	2.67208	8.44985	7.64	58.3696	2.76405	8.74071	8.14	66.2596	2.85307	9.02219
7.15	51.1225	2.67395	8.45577	7.65	58.5225	2.76586	8.74643	8.15	66.4225	2.85482	9.02774
7.16	51.2656	2.67582	8.46168	7.66	58.6756	2.76767	8.75214	8.16	66.5856	2.85657	9.03327
7.17	51.4089	2.67769	8.46759	7.67	58.8289	2.76948	8.75785	8.17	66.7489	2.85832	9.03881
7.18	51.5524	2.67955	8.47349	7.68	58.9824	2.77128	8.76356	8.18	66.9124	2.86007	9.04434
7.19	51.6961	2.68142	8.47939	7.69	59.1361	2.77308	8.76926	8.19	67.0761	2.86182	9.04986
7.20	51.8400	2.68328	8.48528	7.70	59.2900	2.77489	8.77496	8.20	67.2400	2.86356	9.05539
7.21	51.9841	2.68514	8.49117	7.71	59.4441	2.77669	8.78066	8.21	67.4041	2.86531	9.06091
7.22	52.1284	2.68701	8.49706	7.72	59.5984	2.77849	8.78635	8.22	67.5684	2.86705	9.06642
7.23	52.2729	2.68887	8.50294	7.73	59.7529	2.78029	8.79204	8.23	67.7329	2.86880	9.07193
7.24	52.4176	2.69072	8.50882	7.74	59.9076	2.78209	8.79773	8.24	67.8976	2.87054	9.07744
7.25	52.5625	2.69258	8.51469	7.75	60.0625	2.78388	8.80341	8.25	68.0625	2.87228	9.08295
7.26	52.7076	2.69444	8.52056	7.76	60.2176	2.78568	8.80909	8.26	68.2276	2.87402	9.08845
7.27	52.8529	2.69629	8.52643	7.77	60.3729	2.78747	8.81476	8.27	68.3929	2.87576	9.09395
7.28	52.9984	2.69815	8.53229	7.78	60.5284	2.78927	8.82043	8.28	68.5584	2.87750	9.09945
7.29	53.1441	2.70000	8.53815	7.79	60.6841	2.79106	8.82610	8.29	68.7241	2.87924	9.10494
7.30	53.2900	2.70185	8.54400	7.80	60.8400	2.79285	8.83176	8.30	68.8900	2.88097	9.11043
7.31	53.4361	2.70370	8.54985	7.81	60.9961	2.79464	8.83742	8.31	69.0561	2.88271	9.11592
7.32	53.5824	2.70555	8.55570	7.82	61.1524	2.79643	8.84308	8.32	69.2224	2.88444	9.12140
7.33	53.7289	2.70740	8.56154	7.83	61.3089	2.79821	8.84873	8.33	69.3889	2.88617	9.12688
7.34	53.8756	2.70924	8.56738	7.84	61.4656	2.80000	8.85438	8.34	69.5556	2.88791	9.13236
7.35	54.0225	2.71109	8.57321	7.85	61.6225	2.80179	8.86002	8.35	69.7225	2.88964	9.13783
7.36	54.1696	2.71293	8.57904	7.86	61.7796	2.80357	8.86566	8.36	69.8896	2.89137	9.14330
7.37	54.3169	2.71477	8.58487	7.87	61.9369	2.80535	8.87130	8.37	70.0569	2.89310	9.14877
7.38	54.4644	2.71662	8.59069	7.88	62.0944	2.80713	8.87694	8.38	70.2244	2.89482	9.15423
7.39	54.6121	2.71846	8.59651	7.89	62.2521	2.80891	8.88257	8.39	70.3921	2.89655	9.15969
7.40	54.7600	2.72029	8.60233	7.90	62.4100	2.81069	8.88819	8.40	70.5600	2.89828	9.16515
7.41	54.9081	2.72213	8.60814	7.91	62.5681	2.81247	8.89382	8.41	70.7281	2.90000	9.17061
7.42	55.0564	2.72397	8.61394	7.92	62.7264	2.81425	8.89944	8.42	70.8964	2.90172	9.17606
7.43	55.2049	2.72580	8.61974	7.93	62.8849	2.81603	8.90505	8.43	71.0649	2.90345	9.18150
7.44	55.3536	2.72764	8.62554	7.94	63.0436	2.81780	8.91067	8.44	71.2336	2.90517	9.18695
7.45	55.5025	2.72947	8.63134	7.95	63.2025	2.81957	8.91628	8.45	71.4025	2.90689	9.19239
7.46	55.6516	2.73130	8.63713	7.96	63.3616	2.82135	8.92188	8.46	71.5716	2.90861	9.19783
7.47	55.8009	2.73313	8.64292	7.97	63.5209	2.82312	8.92749	8.47	71.7409	2.91033	9.20326
7.48	55.9504	2.73496	8.64870	7.98	63.6804	2.82489	8.93308	8.48	71.9104	2.91204	9.20869
7.49	56.1001	2.73679	8.65448	7.99	63.8401	2.82666	8.93868	8.49	72.0801	2.91376	9.21412
7.50	56.2500	2.73861	8.66025	8.00	64.0000	2.82843	8.94427	8.50	72.2500	2.91548	9.21954

Illustrations: for $n = 8.39$

$$8.39^2 = 70.3921 \qquad 0.839^2 = 0.703921$$

$$83.9^2 = 7039.21 \qquad 0.0839^2 = 0.00703921$$

$$839^2 = 703921 \qquad 0.00839^2 = 0.0000703921$$

$$8390^2 = 70392100$$

Table of Squares and Square Roots *(Continued)*

n	n²	√n	√10n	n	n²	√n	√10n	n	n²	√n	√10n
8.50	72.2500	2.91548	9.21954	**9.00**	81.0000	3.00000	9.48683	**9.50**	90.2500	3.08221	9.74679
8.51	72.4201	2.91719	9.22497	9.01	81.1801	3.00167	0.49210	9.51	90.4401	3.08383	9.75192
8.52	72.5904	2.91890	9.23038	9.02	81.3604	3.00333	9.49737	9.52	90.6304	3.08545	9.75705
8.53	72.7609	2.92062	9.23580	9.03	81.5409	3.00500	9.50263	9.53	90.8209	3.08707	9.76217
8.54	72.9316	2.92233	9.24121	9.04	81.7216	3.00666	9.50789	9.54	91.0116	3.08869	9.76729
8.55	73.1025	2.92404	9.24662	9.05	81.9025	3.00832	9.51315	9.55	91.2025	3.09031	9.77241
8.56	73.2736	2.92575	9.25203	9.06	82.0836	3.00998	9.51840	9.56	91.3936	3.09192	9.77753
8.57	73.4449	2.92746	9.25743	9.07	82.2649	3.01164	9.52365	9.57	91.5849	3.09354	9.78264
8.58	73.6164	2.92916	9.26283	9.08	82.4464	3.01330	9.52890	9.58	91.7764	3.09516	9.78775
8.59	73.7881	2.93087	9.26823	9.09	82.6281	3.01496	9.53415	9.59	91.9681	3.09677	9.79285
8.60	73.9600	2.93258	9.27362	**9.10**	82.8100	3.01662	9.53939	**9.60**	92.1600	3.09839	9.79796
8.61	74.1321	2.93428	9.27901	9.11	82.9921	3.01828	9.54463	9.61	92.3521	3.10000	9.80306
8.62	74.3044	2.93598	9.28440	9.12	83.1744	3.01993	9.54987	9.62	92.5444	3.10161	9.80816
8.63	74.4769	2.93769	9.28978	9.13	83.3569	3.02159	9.55510	9.63	92.7369	3.10322	9.81326
8.64	74.6496	2.93939	9.29516	9.14	83.5396	3.02324	9.56033	9.64	92.9296	3.10483	9.81835
8.65	74.8225	2.94109	9.30054	9.15	83.7225	3.02490	9.56556	9.65	93.1225	3.10644	9.82344
8.66	74.9956	2.94279	9.30591	9.16	83.9056	3.02655	9.57079	9.66	93.3156	3.10805	9.82853
8.67	75.1689	2.94449	9.31128	9.17	84.0889	3.02820	9.57601	9.67	93.5089	3.10966	9.83362
8.68	75.3424	2.94618	9.31665	9.18	84.2724	3.02985	9.58123	9.68	93.7024	3.11127	9.83870
8.69	75.5161	2.94788	9.32202	9.19	84.4561	3.03150	9.58645	9.69	93.8961	3.11288	9.84378
8.70	75.6900	2.94958	9.32738	**9.20**	84.6400	3.03315	9.59166	**9.70**	94.0900	3.11448	9.84886
8.71	75.8641	2.95127	9.33274	9.21	84.8241	3.03480	9.59687	9.71	94.2841	3.11609	9.85393
8.72	76.0384	2.95296	9.33809	9.22	85.0084	3.03645	9.60208	9.72	94.4784	3.11769	9.85901
8.73	76.2129	2.95466	9.34345	9.23	85.1929	3.03809	9.60729	9.73	94.6729	3.11929	9.86408
8.74	76.3876	2.95635	9.34880	9.24	85.3776	3.03974	9.61249	9.74	94.8676	3.12090	9.86914
8.75	76.5625	2.95804	9.35414	9.25	85.5625	3.04138	9.61769	9.75	95.0625	3.12250	9.87421
8.76	76.7376	2.95973	9.35949	9.26	85.7476	3.04302	9.62289	9.76	95.2576	3.12410	9.87927
8.77	76.9129	2.96142	9.36483	9.27	85.9329	3.04467	9.62808	9.77	95.4529	3.12570	9.88433
8.78	77.0884	2.96311	9.37017	9.28	86.1184	3.04631	9.63328	9.78	95.6484	3.12730	9.88939
8.79	77.2641	2.96479	9.37550	9.29	86.3041	3.04795	9.63846	9.79	95.8441	3.12890	9.89444
8.80	77.4400	2.96648	9.38083	**9.30**	86.4900	3.04959	9.64365	**9.80**	96.0400	3.13050	9.89949
8.81	77.6161	2.96816	9.38616	9.31	86.6761	3.05123	9.64883	9.81	96.2361	3.13209	9.90454
8.82	77.7924	2.96985	9.39149	9.32	86.8624	3.05287	9.65401	9.82	96.4324	3.13369	9.90959
8.83	77.9689	2.97153	9.39681	9.33	87.0489	3.05450	9.65919	9.83	96.6289	3.13528	9.91464
8.84	78.1456	2.97321	9.40213	9.34	87.2356	3.05614	9.66437	9.84	96.8256	3.13688	9.91968
8.85	78.3225	2.97489	9.40744	9.35	87.4225	3.05778	9.66954	9.85	97.0225	3.13847	9.92472
8.86	78.4996	2.97658	9.41276	9.36	87.6096	3.05941	9.67471	9.86	97.2196	3.14006	9.92975
8.87	78.6769	2.97825	9.41807	9.37	87.7969	3.06105	9.67988	9.87	97.4169	3.14166	9.93479
8.88	78.8544	2.97993	9.42338	9.38	87.9844	3.06268	9.68504	9.88	97.6144	3.14325	9.93982
8.89	79.0321	2.98161	9.42868	9.39	88.1721	3.06431	9.69020	9.89	97.8121	3.14484	9.94485
8.90	79.2100	2.98329	9.43398	**9.40**	88.3600	3.06594	9.69536	**9.90**	98.0100	3.14643	9.94987
8.91	79.3881	2.98496	9.43928	9.41	88.5481	3.06757	9.70052	9.91	98.2081	3.14802	9.95490
8.92	79.5664	2.98664	9.44458	9.42	88.7364	3.06920	9.70567	9.92	98.4064	3.14960	9.95992
8.93	79.7449	2.98831	9.44987	9.43	88.9249	3.07083	9.71082	9.93	98.6049	3.15119	9.96494
8.94	79.9236	2.98998	9.45516	9.44	89.1136	3.07246	9.71597	9.94	98.8036	3.15278	9.96995
8.95	80.1025	2.99166	9.46044	9.45	89.3025	3.07409	9.72111	9.95	99.0025	3.15436	9.97497
8.96	80.2816	2.99333	9.46573	9.46	89.4916	3.07571	9.72625	9.96	99.2016	3.15595	9.97998
8.97	80.4609	2.99500	9.47101	9.47	89.6809	3.07734	9.73139	9.97	99.4009	3.15753	9.98499
8.98	80.6404	2.99666	9.47629	9.48	89.8704	3.07896	9.73653	9.98	99.6004	3.15911	9.98999
8.99	80.8201	2.99833	9.48156	9.49	90.0601	3.08058	9.74166	9.99	99.8001	3.16070	9.99500
9.00	81.0000	3.00000	9.48683	**9.50**	90.2500	3.08221	9.74679	**10.00**	100.000	3.16228	10.0000

Illustrations: for $n = 8.39$

$$\sqrt{8.39} = 2.89655 \qquad \sqrt{0.839} = 0.915969$$

$$\sqrt{83.9} = 9.15969 \qquad \sqrt{0.0839} = 0.289655$$

$$\sqrt{839} = 28.9655 \qquad \sqrt{0.00839} = 0.0915969$$

$$\sqrt{8390} = 91.5969 \qquad \sqrt{0.000839} = 0.0289655$$

INDEX

Accuracy of numbers, 111
Acute angle, 4
Adjacent angles, 6
Alternate-exterior angles, 9
Alternate-interior angles, 9
Altitude, 33, 34, 81
Angle, 1
  acute, 4
  base, 11, 18
  bisector, 23
  central, 55
  inscribed, 56
  measure of, 2
  obtuse, 4
  right, 2
  straight, 2
Approximate numbers, 111
Arc, 55
  length of, 60
Area, 32, 63, 88
  lateral, 72, 77, 82, 84
  total, 73, 78

Base angle, 11
Base, of parallelogram, 33
  of prism, 70
  of pyramid, 81
  of rectangular solid, 72
  of trapezoid, 18
  of triangle, 34
Calculator, 40
Center, of circle, 54
  of sphere, 87
Central angle, 55
Chord, 55
Circle, 54
  area of, 63
  circumference of, 59
Complement, 8
Complementary angles, 7
Concentric circles, 54
Cone, 83
  right circular, 84
Congruent triangles, 45
Conical surface, 83

Corresponding angles, 9, 44
Corresponding sides, 44
Cube, 71
  total area of, 71
  volume of, 71
Cylinder, 76
  hollow, 78
  right circular, 77
Cylindrical surface, 76

Degrees, 1
Diagonal, 18
Diameter, 54, 87

Edge, of cube, 71
  of polyhedron, 70
  of pyramid, 81
Equilateral triangle, 11

Face, of polyhedron, 70
  of pyramid, 81

Great circle, 87

Height, 33, 34, 81
Hexagon, 11
Hollow cylinder, 78
Hypotenuse, 12, 39

Inscribed angle, 56
Intercepted arc, 56
Isosceles trapezoid, 20
Isosceles triangle, 11, 81

Lateral, area, 72, 77, 82, 84
  edge, 81
  face, 72, 81
Legs of right triangle, 12
Length of arc, 60
Line, 1
Logarithms, 40

Major arc, 55
Measure, of angle, 2
  of arc, 55
Median, 23
Minor arc, 55
Minute, 3

Obtuse angle, 4

Parallel lines, 8
Parallelogram, 16
   altitude of, 33
   area of, 33
   base of, 33
   perimeter of, 27
Pentagon, 11
Perimeter, 24, 82
Perpendicular lines, 2
π, 59
Plane, 1
Point, 1
Polygon, 11
Polyhedron, 70
Precision of numbers, 111
Prism, 70
   base of, 70
   lateral area of, 72
   volume of, 72
Protractor, 2
Pyramid, 81
   base of, 81
   lateral area of, 82
   regular, 81
   volume of, 81
Pythagorean theorem, 39

Quadrilateral, 11, 16
   measure of angles, 18
   perimeter of, 25

Radius, 54, 56, 87
Rectangle, 17
   area of, 32
   perimeter of, 27
Rectangular solid, 70
   total area of, 73
   volume of, 71, 72
Rhombus, 16
   perimeter of, 27
Right angle, 2
Right circular cone, 84
   lateral area of, 84
   volume of, 84
Right circular cylinder, 77
   lateral area of, 77
   total area of, 78
   volume of, 77
Right triangle, 12
Rounding off, 40, 112

Scale drawing, 47
Scalene triangle, 11
Secant, 55
Second, 3
Sector, 64
Side of angle, 1
Significant digits, 40, 112
Similar triangles, 43
Slant height, 82, 84
Slide rule, 40
Sphere, 87
   surface area of, 88
   volume of, 88
Square, 17
   area of, 33
   perimeter of, 27
Square root, 40, 113
Straight angle, 2
Supplement, 8
Supplementary angle, 7

Tangent, 55, 56
Transversal, 8
Trapezoid, 18
   area of, 35
Triangle, 11
   altitude of, 34
   area of, 34
   base of, 34
   equilateral, 11
   isosceles, 11
   measure of angles, 13
   perimeter of, 25
   right, 12
   scalene, 11
Total area, 73, 78

Units of measurement, 42, 94

Vertex, of angle, 1
   of pyramid, 81
Vertical angles, 6
Volume, 70, 77